Bitcoin

Also from EATMS Productions

Books on power, survival, women's autonomy, and the systems shaping modern America.

Nonfiction

Billionaires, Capitalism, and Power

Evil and the Mountain Ungreed
Self Help for American Billionaires
Selfish Steve and the Ivory Tower
Tariffs, Taxes, & Face-Eating Leopards
Ban Billionaires: Fascism Fix

Fascism, Religion, and Cultural Control

Self Help for the Manosphere
Fascism 2025
Fascism & the Perverts & the Greed Virus
Christian Fascism Marriage Book
Tyranny, Table Manners, & Tiramisu

Guides for Women's Autonomy and Protection

How to Survive in Post-America as a Woman
Project 2025 American Drag
4B – Burn, Ban, Boycott, Build
4B OG – So No Go GYN
I'm Glad He's Dead

Analysis of Authoritarian Project 2025

Project 2025: The Blueprint
Project 2025: The List
Project 2025, Christian Dumb Dumbs, & The Republican Agenda
Fascism, Project 2025, & The Pinkprint

Modern Rewrites for Women

Stoic Principles Reimagined
Siddhartha Reimagined
The Prince Reimagined for Women
The Art of War Reimagined for Women
The Jungle Reimagined
The Constitution Reimagined for Women

Machine Learning Series

AI, Bitcoin, Nostr for Women
AI, Safety, & Security for Women
AI, Anxiety, & Health for Women
AI, Kids, & Family Safety for Women
AI, Creativity, & Personal Expression for Women
AI, Independent Work, & Parallel Power for Women

Social Systems Series

Emotional Labor for Women
Household Power for Women
Workplace Power for Women
Medical Bias for Women
Aging Systems for Women
Recovery Systems for Women

Fiction

Dystopian Stories of Resistance and Collapse

Propaganda Paige & the Missing Prosperity
Propaganda Paige & the TIDE Manifesto
Propaganda Paige & the Shadow Cartographers
Propaganda Paige & the Prosperity Alliance
Propaganda Paige & the Shattered Truth
Propaganda Paige & the Rising TIDE
Propaganda Paige & the Last Bastion
Propaganda Paige & the Dawn of Prosperity
Project 2025: Dorian — The Last Men
Project 2025: Boy — A Last Men Novel

Bitcoin for Women

The Autonomy Series
Book 1

by
Mads Duchamp

EATMS
PRODUCTIONS

ISBN: 978-1-966014-45-4

Cover, interior design by: Esme Mees

eatms@pm.me
www.eatms.me

Check out EATMS Underground:
https://tinyurl.com/eatmsNOSTR

Printed in the United States of America.

Money has always been a question of power, not arithmetic.

— Simone de Beauvoir

Table of Contents

Foreword

Money shapes women's lives long before anyone explains how it works. It determines where she can live, how quickly she can leave a bad situation, whether she can help family members, and how much control she has over her own future. Yet most women are taught to treat money as something technical, intimidating, or best left to experts. Confusion is framed as normal. Dependence is framed as safety. Over time, that confusion becomes a quiet form of control, especially in systems that reward compliance and punish deviation without explanation.

Bitcoin enters this landscape not as a promise, but as a contrast. It exposes how much modern money relies on permission, surveillance, and intermediaries that can say yes or no without accountability. For women, those layers matter. Accounts can be frozen. Transfers can be delayed. Rules can change midstream. In high control environments, whether corporate or governmental, financial systems are often designed to remember everything and revisit nothing. Bitcoin does not remove risk or responsibility, but it changes who holds authority.

This book focuses on understanding that shift. It examines how money moves, who controls it, and how those dynamics affect women differently across borders and systems. The goal is not persuasion or prediction. It is to make the structure visible so decisions are informed rather than assumed.

—Mads Duchamp, Winter 2026

~1
Why Money is Never Neutral

The Constraint

Money is often described as neutral. A tool. A system of numbers that works the same for everyone. This framing is comforting, and it is wrong. Money has always functioned as a system of power. It shapes behavior, limits movement, and enforces compliance long before anyone names it as control. For women, money does not simply enable choice. It determines which choices remain possible when conditions tighten.

Modern financial systems appear orderly and objective. Accounts are opened through standardized steps. Transactions are monitored in the name of safety. Rules are enforced through policy rather than discretion. This appearance of neutrality hides where authority sits. Financial systems are built to protect institutions first. Individuals are granted access, but that access is conditional and revocable.

Neutrality fails in moments of urgency. Leaving an unsafe situation. Paying for childcare without notice. Covering medical costs during delays or denials. Helping family members quickly. In these moments, money stops being abstract. It becomes a gatekeeper. Who controls the account matters. How fast money moves matters. Whether transactions can be questioned or halted matters. Women encounter these constraints more often because they are more likely to carry responsibility without holding formal authority.

Financial systems reward predictability. Regular income. Consistent spending. Linear life paths. Deviations are treated as risk. This logic serves large institutions well. It serves lives shaped by interruption poorly. Women's financial histories are more likely to include pauses, part-time work, unpaid caregiving, and informal support. These patterns are common, yet the system treats them as anomalies that require monitoring.

Automation intensifies this imbalance. Algorithms flag behavior that falls outside expected patterns. Sudden withdrawals. Transfers to new recipients. Cross-border payments. Shared accounts supporting multiple people. These flags trigger reviews and holds. Explanations are vague. Appeals are limited. The system records irregularity without recording context. Authority moves in one direction.

For women, the consequences are immediate. She is often expected to stabilize households, absorb shocks, and resolve problems quickly. When money is delayed, the impact is not theoretical. Rent is late. Care arrangements collapse. Travel becomes impossible. Waiting itself becomes a form of pressure. Compliance is rewarded. Urgency is punished.

Money also operates as surveillance. Every transaction creates a permanent record. That record can be shared and reviewed long after the moment has passed. Financial privacy has narrowed quietly through compliance frameworks framed as protection. Anti-fraud and anti-laundering rules expand monitoring while reducing discretion. Women relying on informal support networks or mutual aid are more exposed to scrutiny.

Surveillance changes behavior before enforcement occurs. When women know transactions can be questioned later, they adapt in advance. They avoid direct transfers. They keep balances low. They delay action until it feels safer. This is not paranoia. It is learned caution. The system does not need to intervene often if the possibility of intervention shapes behavior.

Money determines mobility. Access to funds is access to movement. Many women discover the limits of financial freedom only when they try to leave or relocate. Cards fail. Transfers stall. Accounts are frozen during verification. Identity checks escalate when urgency increases. The system reads urgency as risk.

These constraints are not accidental. Financial systems are designed to slow movement. Withdrawal limits, transaction caps, and reviews are framed as safeguards. They function equally as barriers. For women in high control environments, restricted access to money reinforces dependence. Choice becomes conditional.

Even in stable economies, money shapes safety. A woman with independent access to funds can leave a hostile job. A woman whose finances are shared or monitored cannot act quietly. Financial independence is praised in theory. In practice, it is easily undermined by systems that allow use but retain control.

The language of finance obscures this imbalance. Accounts are "frozen," not seized. Transactions are "delayed," not denied. Reviews are "routine," not accusatory. These terms soften the reality that individuals have no reciprocal authority. Institutions are not subject to equivalent interruption.

Women are often told that financial literacy will solve this. Learn the rules. Budget better. Plan ahead. These suggestions assume the system responds to preparation. In reality, preparation works only within narrow bounds. No amount of literacy prevents a freeze. No plan eliminates sudden rule changes. Knowledge does not equal power.

Credit deepens constraint. Access to credit is framed as empowerment. It often binds women to future obligations that reduce flexibility. Debt assumes stability. Missed payments compound quickly. Credit scoring systems penalize non-linear work histories. Women who leave the workforce for caregiving are marked as higher risk. The system records absence without recording cause.

As systems tighten, these pressures intensify. Regulation expands. Institutions become more cautious. Automated enforcement replaces human judgment. Appeals shrink. What once required explanation becomes an algorithmic decision. Financial control is rarely announced. It is applied through existing compliance mechanisms that appear neutral.

Money becomes a sorting tool. Who is compliant. Who is risky. Who requires oversight. This sorting remains invisible until it disrupts daily life. Women experience it as friction rather than force. Extra paperwork. Repeated verification. Requests for justification. Each step reinforces that access is conditional.

The burden is uneven. Women with fewer resources are more exposed. Single mothers, migrant women, caregivers, women leaving unsafe situations all interact with financial systems under pressure. Their transactions look irregular because their lives are irregular. The system does not adjust. It flags. This creates a cycle. Instability increases scrutiny. Scrutiny increases instability. Women are expected to remain patient while delays threaten safety and autonomy. Emotional labor becomes part of financial survival. She must explain, reassure, and wait.

Money is not neutral because it always answers to someone. It functions as an infrastructure of permission. It determines who can act quickly, who must ask, and who must wait. For women everywhere, safety, mobility, and choice are filtered through systems that were not built to prioritize their autonomy. Understanding this constraint is the necessary starting point.

The Shift

Recognizing that money is not neutral changes the frame. It moves the conversation away from personal failure and toward structure. Once women see that financial systems operate as permissioned infrastructure rather than passive tools, a different question becomes possible. The question is no longer how to behave better inside the system, but where authority actually sits and whether it can be shifted.

The shift begins with understanding money as design. Every financial system encodes values. Speed or delay. Privacy or visibility. Reversibility or finality. Central control or distributed control. These choices are not abstract. They determine who decides, who waits, and who explains. When women learn to read money systems this way, neutrality disappears and leverage becomes visible.

In traditional systems, authority rests with intermediaries. Banks, payment processors, regulators, and platforms sit between individuals and their money. Access is granted conditionally. Rules can change without consent. Reviews can happen without explanation. This arrangement is treated as normal because it has been normalized over time. The shift is not to deny that intermediaries exist, but to recognize that their control is structural, not inevitable.

For women, this recognition restores agency. It clarifies that many financial constraints are not personal shortcomings. They are outcomes of systems optimized for oversight rather than autonomy. When responsibility and authority are misaligned, frustration follows. When that misalignment is named, strategy becomes possible.

The shift also reframes financial independence. Independence is often presented as income alone. Earn more. Save more. Invest wisely. These goals matter, but they are incomplete. Independence is also about control. Who can stop a transaction. Who can delay it. Who can demand explanation. A woman who earns money but cannot move it freely is not fully independent. Seeing this distinction sharpens decision-making.

Understanding money as infrastructure encourages optionality. Instead of assuming a single system must serve every purpose, women can begin to ask which tools are appropriate for which needs. Some systems are designed for stability and integration. Others are designed for resilience and exit. The shift is not rejection. It is discernment.

This perspective is especially important in tightening environments. When systems become more restrictive, dependence becomes more costly. Women who understand where financial control lives are better positioned to reduce single points of failure. They are less likely to assume that compliance guarantees safety. They plan with interruption in mind, not as paranoia, but as realism.

The shift also changes how risk is understood. In traditional framing, deviation is risky and compliance is safe. In reality, dependence carries its own risk. A system that can freeze funds without recourse concentrates danger. Diversifying control reduces that exposure. When women see dependence itself as a vulnerability, risk assessment becomes more balanced.

This does not require technical expertise. It requires conceptual clarity. Money systems are not moral. They are functional. They do what they are designed to do. When women stop expecting neutrality, they stop being surprised by outcomes. Surprise is costly. Anticipation restores steadiness.

The shift also returns time to women. Traditional systems extract time through delays, reviews, and explanations. Waiting is treated as neutral, but it is not. Waiting consumes attention and energy. When women understand that delay is a feature rather than an accident, they stop interpreting it as a personal problem to solve through patience alone. They begin to value speed and finality where it matters.

Another part of the shift is separating legitimacy from permission. Traditional money systems equate legitimacy with approval. If a transaction is allowed, it is valid. If it is blocked, it is suspect. This framing internalizes authority. Women begin to question their own needs rather than the system's design. The shift is recognizing that permission does not equal morality or necessity. It is simply a rule applied by an institution.

This recognition matters in moments of urgency. A blocked transaction does not mean the need is invalid. A delay does not mean the action was wrong. When women detach legitimacy from permission, shame loses its grip. Decision-making becomes clearer and calmer.

The shift also reframes privacy. Financial privacy is often dismissed as secrecy or avoidance. In reality, privacy is about proportionality. Not every transaction requires permanent record. Not every act of support needs oversight. When women understand privacy as a condition for safety rather than concealment, they begin to value systems that minimize exposure.

This perspective aligns with lived experience. Women know that visibility carries uneven consequences. They already manage disclosure carefully in other areas of life. Extending that awareness to money is not radical. It is consistent.

Understanding money as a power system also clarifies why alternatives exist at all. New financial tools do not emerge because the old ones are broken. They emerge because the old ones are doing exactly what they were designed to do, and that design does not serve everyone equally. The shift is not believing that alternatives are perfect. It is understanding what problem they are responding to. For women, this clarity is stabilizing. It removes the pressure to adopt identities or ideologies. One does not need to believe in any system to benefit from understanding it. Literacy restores choice. Choice restores calm.

The shift also encourages boundary-setting. Women are often expected to absorb financial friction quietly. Extra forms. Extra steps. Extra explanations. Seeing these demands as structural rather than personal makes refusal possible. Not every burden must be accepted as normal. Some can be planned around. Others can be avoided.

In authoritarian contexts, this shift becomes essential. Control tightens through existing systems long before new rules are announced. Women who understand money as infrastructure are less likely to be caught off guard. They are not immune to pressure, but they are less disoriented by it. Disorientation is one of the most effective tools of control. The shift does not promise safety. No financial system can. What it offers is clarity. Clarity about where power lives. Clarity about tradeoffs. Clarity about what is being exchanged when convenience is chosen over control.

This chapter does not ask women to abandon familiar systems. It asks them to see those systems clearly. Once neutrality is abandoned, authority becomes visible. Once authority is visible, it can be negotiated, diversified, or reduced.

Money stops being a moral scorecard and becomes what it has always been. A mechanism. One that can be used, questioned, and supplemented. The shift is not ideological. It is practical. When women understand that money is never neutral, they stop waiting for it to be fair. They stop expecting it to protect them. They begin to decide deliberately how much control they are willing to give and where. That decision is the beginning of autonomy.

~2
The Problem with Traditional Banking

The Constraint

Traditional banking is often presented as a public good. A stable system designed to keep money safe, transactions orderly, and economies functioning. Banks describe themselves as neutral intermediaries, serving everyone equally under the same rules. In practice, traditional banking works extremely well for institutions and governments. It works far less well for women whose lives do not fit the assumptions those systems are built on.

Banks are optimized for scale, compliance, and risk management. Their primary responsibility is not to individual customers, but to regulators, shareholders, and the stability of the system itself. Individuals interact with banks through standardized products and procedures designed to reduce institutional liability. This structure prioritizes uniformity over context. For women, whose financial lives are more likely to be complex, interrupted, or relational, that uniformity becomes a source of friction.

Institutional clients benefit from predictability. Large, regular transactions. Clear documentation. Stable organizational identities. Governments benefit from traceability. Banks serve both by maintaining detailed records, enforcing reporting requirements, and restricting movement when activity appears unusual.

These functions are framed as safeguards. They are also mechanisms of control. What protects institutions often constrains individuals.

Women encounter this constraint when their financial needs do not align with institutional expectations. Irregular income. Shared resources. Informal caregiving arrangements. Supporting multiple people across households or borders. These realities are common, yet banking systems treat them as risk signals rather than normal life patterns. Accounts are flagged. Transfers are delayed. Access is questioned.

Traditional banks rely heavily on automated risk assessment. Algorithms evaluate behavior against expected norms. Sudden changes trigger scrutiny. Cash withdrawals. Transfers to new recipients. International payments. The system does not distinguish between danger and necessity. It does not ask why. It flags first and explains later, if at all.

For women, this creates a recurring vulnerability. She is often responsible for moving money quickly to meet immediate needs. When banking systems slow or stop those movements, the consequences are personal. Rent is due. Care arrangements depend on timing. Travel may be necessary for safety or family obligation. Delays that appear minor on paper can have cascading effects in real life.

Banks also assume individual financial autonomy that many women do not fully possess. Joint accounts, shared credit lines, and household-based products obscure unequal power dynamics. The bank recognizes the account structure, not the relationships inside it. When conflict arises, formal ownership determines access. Contribution, caregiving labor, and dependency are invisible to the system.

This becomes especially dangerous when women need to act independently. Leaving a controlling partner. Separating finances quietly. Protecting resources from interference. Traditional banking is poorly suited to discreet action. Activity is logged. Notifications are sent. Changes are visible. The system prioritizes transparency over safety. For women in high-control environments, this transparency can expose plans before they are secure.

Banks are also deeply intertwined with government policy. Regulations require cooperation with surveillance and enforcement frameworks. Financial institutions are expected to monitor customers and report suspicious activity. These obligations align banks with state power, not individual autonomy. When governments tighten controls, banks become the enforcement layer. This is not an abuse of the system. It is its design.

For women, this alignment matters. Policies applied broadly do not affect everyone equally. Reporting thresholds, verification requirements, and compliance checks are experienced as friction by those with less margin. Women managing household finances, caregiving, or migration often operate close to those thresholds. Each new requirement increases vulnerability.

Access to banking is often framed as inclusion. Having an account is treated as empowerment. In reality, access without control is limited empowerment. Banks can close accounts with little notice. They can restrict services based on policy changes. Appeals are slow and opaque. Individuals have few remedies. The power imbalance is structural.

Credit products illustrate this clearly. Loans and credit cards are offered as tools for flexibility. They bind women to future obligations that assume stability. Interest compounds. Missed payments escalate quickly. Credit scoring systems penalize non-linear employment histories and caregiving interruptions. The system records absence without recording reason.

Banks also extract time. Documentation requirements. Verification processes. Customer service loops. Women are expected to navigate these burdens calmly while maintaining daily responsibilities. Time spent resolving banking issues is time taken from work, caregiving, and recovery. This labor is unpaid and invisible.

When systems tighten, banking becomes more restrictive. Compliance expands. Automation increases. Human discretion disappears. What once required conversation becomes a policy outcome. Accounts are frozen pending review. Transactions are reversed without explanation. The customer is expected to wait. Traditional banking's failure is not incompetence. It is misalignment. The system is designed to serve institutional priorities. Stability, oversight, and traceability matter more than individual flexibility. For governments and corporations, this works. For women navigating complex lives, it often does not.

This failure is reinforced by the language of responsibility. Women are told to keep better records, plan further ahead, and avoid risk. These instructions assume the system will respond proportionally. It does not. Uniform rules produce unequal impact. Preparation does not eliminate vulnerability.

The result is a quiet erosion of autonomy. Women adapt by self-limiting. They avoid moving money. They delay action. They accept dependence. This adaptation is rational within constraint, but it narrows options over time.

Traditional banking is not neutral infrastructure. It is a system of permission, surveillance, and enforcement. It works well for institutions that benefit from control and visibility. It often fails women because it was not built to account for lives shaped by care, interruption, and urgency. Naming this constraint is the first step toward understanding why alternatives matter at all.

The Shift

The shift begins when women stop treating traditional banking as a protective institution and start treating it as infrastructure. Infrastructure is not moral. It does not adapt to individual need. It performs specific functions reliably and fails predictably outside those functions. Once banking is understood this way, the relationship changes. Confusion gives way to orientation. Expectations narrow. Surprise loses its power.

Traditional banking excels at institutional coordination. It integrates with employers, tax authorities, courts, and regulatory bodies. It creates durable records, standardizes behavior, and allows governments to monitor economic activity at scale. These features are not accidental. They are the reason banks exist in their current form. The shift is recognizing that these same features make banks poorly suited for discretion, urgency, or personal safety.

For women, this recognition restores agency. Many banking disruptions feel destabilizing because they violate an unspoken assumption of care. When an account is frozen or a transfer delayed, it feels like a personal failure or a system malfunction. When the same event is understood as routine enforcement behavior, the emotional impact changes. The system is not responding to her situation. It is executing policy. Clarity replaces self-blame.

This shift also reframes trust. Trust is often confused with longevity. A long relationship with a bank feels secure because it is familiar, not because interests are aligned. Banks are structurally aligned with regulators, compliance thresholds, and institutional risk management. They are not aligned with individual autonomy. Knowing this allows women to place trust selectively rather than broadly, reducing the risk of overreliance.

Boundaries follow naturally. Banks are reliable for predictable, visible transactions that can tolerate delay and review. They are unreliable in moments requiring speed, discretion, or independence. Seeing this distinction allows women to plan around known fault lines instead of assuming universal functionality. Planning replaces hope. Hope is fragile in systems that do not reciprocate it.

The shift also sharpens the meaning of control. Control is not whose name appears on an account. It is who can interrupt access, reverse movement, or demand explanation. Traditional banking concentrates that authority within the institution. Once this is acknowledged, women can decide intentionally which resources must remain immediately accessible and which can tolerate institutional oversight.

This distinction becomes more important as systems tighten. Compliance rarely contracts. It expands. Automation replaces discretion. Reviews become opaque. Appeals slow or disappear. Women who understand this are less likely to interpret new restrictions as temporary glitches. They recognize enforcement patterns and adjust expectations accordingly. Orientation replaces shock.

Risk looks different through this lens. Compliance is often framed as safety. In reality, dependence concentrates risk. When all resources pass through a single institutional channel, a single freeze can destabilize everything at once. Reducing single points of failure becomes a practical strategy, not an ideological stance.

Women already apply this logic elsewhere. Backup childcare plans. Redundant support networks. Contingency arrangements for caregiving and emergencies. Applying the same reasoning to finances reflects lived experience. It is not radical. It is consistent with how women already manage vulnerability.

Another aspect of the shift is emotional. Banking failures often trigger shame. Why was the account flagged. Why did this happen now. Reframing these events as predictable outcomes of system design removes that burden. Emotional energy can be redirected toward response rather than explanation. Calm returns when meaning is clarified.

The shift also separates legitimacy from permission. Traditional banking equates approval with correctness. If a transaction is allowed, it is legitimate. If it is blocked, it is suspect. This framing internalizes institutional judgment. Women begin questioning their own needs rather than the system's constraints. Recognizing that permission reflects policy, not morality, restores clarity and reduces paralysis.

This distinction matters most in moments of urgency. A blocked transfer does not invalidate the need it was meant to meet. A frozen account does not imply wrongdoing. When legitimacy is detached from approval, women can act based on reality rather than fear of interpretation.

Privacy is also reframed. Banks require maximum visibility because visibility supports oversight and enforcement. That visibility does not serve personal safety equally. Women already practice selective disclosure in professional, medical, and social contexts. Extending that practice to finances is not secrecy. It is proportionality. Privacy here is about scale and permanence. Not every transaction requires a permanent record. Not every act of support needs institutional exposure. When women stop assuming that full transparency is inherently virtuous, they regain discretion without abandoning responsibility.

Understanding traditional banking's limits clarifies why parallel systems exist at all. Alternatives do not arise because banks are broken. They arise because banks are effective at serving priorities that do not meet every need. The shift is not believing alternatives are perfect. It is understanding the constraint they are responding to.

This removes pressure to conform. Women do not need to adopt identities or beliefs to make use of multiple systems. Different tools can serve different purposes. Integration for stability. Separation for resilience. Flexibility replaces loyalty. Loyalty is rarely rewarded by institutions designed for scale rather than reciprocity. The shift also encourages strategic engagement. Women can continue using banks for predictable, visible functions while protecting resources that require speed or autonomy elsewhere. This is not avoidance. It is segmentation. Segmentation reduces exposure without demanding withdrawal.

In authoritarian environments, this shift becomes essential. Control rarely arrives announced. It expands through existing institutions that already manage money, identity, and compliance. Banks become enforcement points because the infrastructure already exists. Women who understand this are not immune to constraint, but they are less disoriented by it. Banks provide integration and convenience in exchange for control and visibility. Once that exchange is visible, women can decide deliberately when it is acceptable and when it is not.

This chapter does not argue for abandoning traditional banking. It argues for removing it from a pedestal it was never meant to occupy. Banks are institutional actors with specific incentives. Seeing them clearly allows engagement without illusion.

The shift is practical, not ideological. It restores agency by aligning expectations with reality. When banking is understood as infrastructure rather than protection, dependence loosens and options expand. That repositioning is the shift.

~3
When Money is Watched and Controlled

The Constraint

Financial surveillance is often described as a safety measure. Monitoring is framed as protection against fraud, crime, and instability. Limits are described as safeguards. Freezes are presented as temporary precautions. Delays are treated as neutral inconveniences. This language softens what is actually happening. Modern financial systems do not merely observe money. They watch behavior, classify it, and intervene without consent.

Surveillance begins quietly. Every transaction is recorded by default. Purchases, transfers, locations, timing, frequency. These data points accumulate into profiles that outlast the moment they were created for. Financial records are not contextual. They do not explain why money moved. They only show that it did. Over time, this creates a permanent behavioral archive that can be searched, shared, and reinterpreted without the participation of the person it represents.

For women, this archive carries uneven risk. Women are more likely to use money relationally. Supporting family members. Managing household expenses. Covering gaps created by caregiving or illness. Sending small amounts frequently. Sharing accounts. These patterns are normal, but they do not look normal to automated systems trained on institutional behavior.

What is common in women's lives often appears irregular in surveillance models.

Financial monitoring systems rely on pattern recognition. They are designed to detect deviation from expected norms. Sudden changes trigger attention. Increased withdrawals. New recipients. Cross-border activity. Account access from different locations. These signals initiate reviews automatically. There is rarely a human assessment at the start. The system intervenes first and asks questions later, if it asks at all.

When money is watched this way, interpretation replaces intent. The system does not know whether a transfer was made out of necessity, care, or urgency. It only sees deviation. That deviation is treated as potential risk. Women experience the result as friction. A transaction fails. An account is flagged. Access is paused. The reason is vague or unavailable. The effect is immediate.

Freezes are one of the most disruptive tools in financial surveillance. An account freeze halts access entirely. Payments stop. Cards fail. Automatic withdrawals bounce. The freeze is often described as temporary, but the duration is unpredictable. Resolution requires documentation, explanation, and time. For women managing households or caregiving responsibilities, time is rarely neutral. Delays can destabilize everything.

Limits operate similarly, but more quietly. Withdrawal caps. Transfer ceilings. Daily spending limits. These restrictions are framed as protective. In practice, they constrain responsiveness. A woman facing an emergency may have money in her account but be unable to access it quickly enough. The system does not adjust for urgency. It enforces limits uniformly, regardless of consequence.

Delays are perhaps the most normalized form of control. Reviews take days or weeks. Verifications require repeated submissions. Customer service loops redirect responsibility. The delay itself becomes a barrier. Women are expected to wait calmly while real-world obligations continue. Rent does not pause. Care needs do not pause. Safety does not pause.

Surveillance also changes behavior before intervention occurs. When women know transactions may be flagged, they adapt in advance. They avoid moving money directly. They break transfers into smaller amounts. They keep balances low. They delay action until it feels safer. This self-regulation is not irrational. It is a response to uncertainty. The system does not need to freeze often if the possibility of freezing shapes behavior.

Financial surveillance also intersects with identity verification. Names, addresses, documents, and biometrics are increasingly required to access funds. Any inconsistency triggers review. Women who have changed names, relocated frequently, crossed borders, or exited relationships encounter higher friction. The system assumes continuity. Women's lives often involve transition. Each transition increases exposure to scrutiny.

These verification processes are framed as objective checks. In practice, they impose unequal burdens. Producing documents requires time, stability, and administrative access. Women managing care, displacement, or recovery are less likely to have records immediately available. The system does not pause for reconstruction. Access is withheld until proof is produced, even when the need for funds is immediate.

Surveillance also collapses financial past, present, and future into a single risk profile. A transaction made years earlier can be reinterpreted under new rules without warning. Behavior that was once acceptable may later be flagged retroactively. Women who adapt over time, shift roles, or respond to changing circumstances carry records that never fully reset. The system remembers behavior but does not recognize growth, context, or explanation. Permanence becomes another layer of pressure.

When surveillance expands, appeals contract. Decisions are automated. Explanations are minimal. Accountability is diffuse. No single person is responsible for the outcome. The system enforces policy without discretion. Women are left navigating opaque processes while their resources are inaccessible, expected to prove legitimacy after access has already been removed.

This control is rarely framed as punishment. It is framed as precaution. That framing obscures power. Precaution applied without recourse becomes coercive. When access to money can be interrupted without explanation, compliance becomes safer than clarity.

Women learn to avoid actions that might draw attention, even when those actions are necessary.

Surveillance also magnifies existing inequalities. Women with fewer resources are more exposed. A small freeze can cascade into eviction, missed care, or loss of employment. Those with buffers can absorb delays. Those without cannot. The system treats these outcomes as external. It does not account for disproportionate impact.

In high-control environments, surveillance becomes a tool of enforcement. Financial monitoring is used to discourage movement, support, and dissent. This does not require explicit bans. It requires uncertainty. When women do not know what will trigger intervention, they limit themselves. Fear enters financial decision-making long before enforcement occurs.

Money that is constantly watched is not neutral. It disciplines behavior. It rewards predictability and punishes urgency. It favors those whose lives align with institutional norms and constrains those whose lives do not. For women, whose financial lives are more likely to be relational, interrupted, and urgent, surveillance is not evenly experienced.

The constraint is not mere inconvenience. It is the loss of timing, discretion, and agency. When money is watched and controlled, women are forced to navigate systems that treat care as anomaly and urgency as risk. Understanding this constraint is necessary before any shift is possible.

The Shift

The shift begins with recognizing that surveillance is not accidental. It is a design choice. Financial systems are built to observe, record, and intervene because those functions serve institutional priorities. Once women understand that monitoring, limits, freezes, and delays are features rather than failures, the relationship to money changes. Surprise fades. Confusion recedes. Orientation replaces shock.

Seeing surveillance clearly reframes responsibility. When access is restricted, the instinct is often to search for personal error. Did I move money the wrong way. Did I trigger something unintentionally. This reflex internalizes blame. The shift is recognizing that intervention does not require wrongdoing. It requires deviation. When deviation is normal, blame loses its usefulness.

This clarity restores steadiness. Women stop interpreting surveillance events as personal judgment and start treating them as predictable outcomes of system design. That mental shift alone reduces panic. Panic narrows options. Orientation expands them.

The shift also changes how women evaluate safety. Surveillance is often sold as protection. In practice, protection is conditional. Systems protect institutions first. Individual safety is secondary and uneven. When women stop equating monitoring with care, they can assess risk more realistically. Visibility is no longer assumed to be beneficial. It becomes a tradeoff.

Understanding surveillance as infrastructure allows women to choose when exposure is acceptable and when it is not. Some transactions benefit from visibility. Others require discretion. The shift is not secrecy. It is proportionality. Not all actions deserve the same level of observation or permanence.

This perspective also changes how timing is valued. Surveillance systems extract time through delays, reviews, and verifications. Waiting is framed as neutral. It is not. Waiting consumes energy and destabilizes planning. When women recognize delay as a control mechanism rather than a logistical inconvenience, they stop treating patience as the solution. They begin valuing systems and strategies that preserve timing. The shift reframes freezes as signals, not anomalies. A freeze reveals where authority resides. It shows who decides, who explains, and who waits. Once this is visible, women can stop expecting recourse where none exists. Expectation management is not resignation. It is preparation.

Limits and caps are also reinterpreted. Rather than assuming limits exist for individual safety, women see them as population-level controls. They are designed to manage risk at scale, not to accommodate urgency. This understanding allows women to plan around limits instead of being trapped by them.

The shift also loosens fear around record permanence. Financial records are durable. They do not forget. But understanding permanence changes how it is handled. Women stop assuming that good behavior guarantees safety. They stop believing that consistency alone prevents scrutiny. When permanence is accepted as a condition rather than a deterrent, decisions become more grounded.

This clarity also separates morality from compliance. Surveillance systems treat rule adherence as virtue. Deviations are treated as suspicion. The shift is recognizing that necessity often conflicts with compliance. Supporting family. Leaving unsafe environments. Responding to crisis. These needs do not disappear because a system is uncomfortable with them. When morality is detached from permission, shame loses its grip. A blocked transaction no longer feels like moral failure. It feels like friction. Friction can be planned for. Shame cannot.

The shift also changes how women respond to identity verification. Verification assumes stability. Women's lives often involve transition. Recognizing this mismatch prevents internalization of blame. A name change is not suspicious. A relocation is not a risk marker. They only appear that way to systems built for continuity.

This awareness encourages strategic documentation and pacing rather than reactive compliance. Women stop scrambling to justify themselves after access is removed. They begin anticipating where friction is likely to occur. Anticipation reduces vulnerability even when constraints remain.

The shift also reframes self-regulation. Many women already limit their behavior to avoid attention. They split transfers. Delay action. Keep balances low. Seen without context, this looks like fear. Seen structurally, it is adaptation. The shift is not eliminating adaptation. It is making it conscious rather than automatic.

When adaptation is conscious, it can be adjusted. Women regain the ability to decide when constraint is worth accepting and when it is not. Automatic self-limiting narrows life. Deliberate choice preserves agency.

Surveillance also loses some of its power once named. Systems that rely on uncertainty function best when people do not understand them. When rules are opaque, fear fills the gap. Clarity weakens that fear. Even when control remains, disorientation does not.

The shift also restores sequencing. Surveillance collapses thinking, acting, and justifying into a single moment of exposure. Women are expected to act correctly, document perfectly, and explain immediately. Recognizing this collapse allows women to rebuild sequence. Action can be separated from explanation. Urgency can be addressed before compliance. This does not remove constraint, but it reduces internal pressure to resolve everything at once.

Finally, the shift reframes vigilance. Vigilance is often treated as paranoia when exercised by individuals, while being normalized when exercised by institutions. Women are encouraged to trust systems that do not trust them. Naming this imbalance changes how vigilance is held. It becomes situational rather than constant. Attention is focused where risk is real, not where fear is implied. That redistribution of attention preserves energy and reduces exhaustion.

In authoritarian environments, this shift becomes essential. Surveillance expands quietly through financial systems long before political language changes. Monitoring intensifies. Enforcement becomes routine. Women who understand this are less likely to mistake escalation for anomaly. They are not immune to pressure, but they are less destabilized by it. The shift does not promise freedom from surveillance. No modern financial system offers that. What it offers is literacy. Literacy restores choice. Choice restores calm. Calm preserves decision-making under pressure.

This chapter does not argue for resisting surveillance directly. It argues for seeing it accurately. When women understand how and why money is watched and controlled, they stop expecting neutrality. They stop assuming fairness. They begin deciding deliberately how much exposure they are willing to tolerate and where.

Money under surveillance disciplines behavior. The shift is refusing to let that discipline become internal law. Women remain constrained, but they are no longer confused about why. That clarity is the beginning of leverage. When money is watched, authority is hidden. When it is named, authority becomes visible. Visibility does not eliminate control, but it changes the relationship to it. That change is the shift.

~4
Bitcoin is a Different Kind of Money

The Constraint

By the time most women hear about Bitcoin, it is already framed incorrectly. It is presented as speculation, rebellion, or technical obsession. Headlines focus on price swings. Critics dismiss it as impractical or dangerous. Advocates promise transformation. None of these frames explain what actually matters. Bitcoin is not primarily an asset or an ideology. It is a response to a structural constraint embedded in traditional money. That constraint is permission.

Traditional money is permissioned at every layer. Accounts exist because institutions allow them to exist. Transfers move because intermediaries approve them. Access continues because rules are followed. Surveillance, freezes, limits, and delays are not malfunctions. They are expressions of who controls the system. Women encounter these controls not as theory, but as friction that shows up precisely when timing, discretion, or autonomy matters most.

Bitcoin emerges from this environment as a different kind of system, not because it is better in every way, but because it removes specific points of control. Understanding that difference requires stepping away from familiar assumptions about money. It requires seeing money not as a product offered by institutions, but as infrastructure shaped by design choices.

Traditional money relies on trusted intermediaries. Banks, payment processors, clearinghouses, and regulators sit between people and their funds. These intermediaries maintain ledgers, verify identity, reverse transactions, and enforce policy. Trust is centralized. Authority flows downward. The system works efficiently for large institutions and governments because it is designed to.

That same design creates vulnerability for individuals. When authority is centralized, access is conditional. When access is conditional, compliance becomes necessary. When compliance is automated, discretion disappears. Women experience this as loss of timing, loss of privacy, and loss of quiet decision-making.

Bitcoin removes intermediaries from the core of the system. It replaces institutional trust with a distributed ledger that anyone can verify and no single entity controls. This is the structural difference that matters. Bitcoin does not require permission to exist. It does not require approval to transact. It does not rely on an institution to maintain the ledger.

This does not mean Bitcoin eliminates risk. It does not mean it is private by default. It does not mean it is immune to surveillance or misuse. It means that the locus of control is different. Control over the system is not held by a single authority that can intervene at will. Traditional money systems are built to be reversible. Transactions can be undone. Accounts can be restored. Errors can be corrected. This reversibility is often presented as protection. In practice, it also enables control. The ability to reverse implies the authority to judge. Someone decides which reversals are allowed and which are not.

Bitcoin transactions are final. Once confirmed, they cannot be reversed by appeal or policy change. This finality is often criticized as dangerous. It is dangerous in some contexts. It also removes a layer of control that exists in traditional systems. There is no institution that can freeze a transaction after the fact.

For women, this distinction matters less as a feature and more as a boundary. Finality changes who must be trusted. It shifts responsibility outward from institutions and inward toward the individual. That shift is not inherently empowering. It requires understanding and care. But it is structurally different from systems that retain the power to intervene indefinitely.

Traditional money systems assume stable identity. Accounts are tied to names, documents, and continuous records. Changes trigger scrutiny. Bitcoin does not require identity at the protocol level. It recognizes cryptographic keys, not personal histories. This does not make it anonymous in practice, but it does remove identity enforcement from the base layer. This difference is often misunderstood. Bitcoin does not promise invisibility. It does not erase past records. It does, however, separate identity from access in a way traditional systems do not. Access is controlled by possession of keys, not by approval from an institution.

In traditional systems, money is entangled with jurisdiction. Borders determine what is allowed, delayed, or blocked. Cross-border movement triggers heightened surveillance and restriction. Bitcoin operates on a global network that does not recognize national borders at the protocol level. This does not eliminate legal consequences, but it removes border enforcement from the mechanism of transfer itself.

This matters because borders are where many women experience the sharpest financial constraints. Supporting family abroad. Leaving unsafe situations. Moving resources quickly across jurisdictions. Traditional systems treat these actions as risk signals. Bitcoin treats them as transactions.

Bitcoin is also structurally limited. Its supply is fixed by design. This is often framed as an economic argument. Structurally, it means the rules governing issuance cannot be changed by policy decision. No committee can expand supply in response to pressure. This removes a tool governments and institutions routinely use to manage systems. For women, the relevance is indirect. Inflation and devaluation disproportionately affect those with less margin. Savings erode quietly. Planning becomes harder. Bitcoin's fixed supply does not solve this problem universally, but it represents a system where certain rules are not subject to discretionary change.

Traditional money systems evolve through policy. Rules change. Thresholds shift. Reporting expands. What was allowed yesterday may be flagged tomorrow. Bitcoin's protocol changes slowly and visibly, requiring broad consensus. This does not make it static or perfect. It makes change harder to impose unilaterally. This constraint on change is often criticized as rigidity. It is rigidity by design. In traditional systems, flexibility serves those with authority. In Bitcoin, rigidity limits authority. That tradeoff is central to understanding why Bitcoin exists at all.

For women accustomed to systems that promise convenience while extracting control, Bitcoin's design is revealing. It does not pretend to protect. It does not offer customer service. It does not reverse mistakes. What it offers instead is a system that does not watch behavior in the same way, does not intervene arbitrarily, and does not depend on institutional permission to function. This difference is not about belief. It is about architecture. Bitcoin treats money as a protocol rather than a product. Protocols define rules and then step back. Products manage users continuously. Traditional money is a product. Bitcoin is a protocol.

Understanding this distinction removes hype. Bitcoin is not a miracle. It introduces new risks while removing others. It shifts responsibility. It trades convenience for autonomy in specific ways. The constraint it responds to is not corruption or greed. It is centralized control over money as infrastructure.

Before any evaluation of usefulness, this structural difference must be understood. Bitcoin exists because traditional money systems are built around permission, surveillance, and intervention. Bitcoin removes some of those levers, not to create freedom, but to change who holds power. That difference is the starting point.

The Shift

The shift begins by moving Bitcoin out of the realm of belief and into the realm of structure. Bitcoin does not need to be admired, trusted, or adopted wholesale to be understood. It only needs to be seen clearly. When women approach Bitcoin as infrastructure rather than ideology, the noise falls away. What remains is a system that rearranges where authority sits.

Traditional money places authority with intermediaries. Banks decide access. Governments shape rules. Platforms enforce compliance. Individuals operate inside those boundaries. Bitcoin rearranges that arrangement by removing the intermediary from the core transaction. This does not eliminate power. It redistributes it. The shift is recognizing that redistribution and understanding what it changes and what it does not.

Bitcoin's most significant shift is that it replaces permission with possession. Access is controlled by cryptographic keys rather than institutional approval. There is no application process at the protocol level. No account to be opened. No authority that can revoke access once keys are held. This does not guarantee safety or success. It does change who decides.

For women accustomed to systems where access can be paused or removed without explanation, this difference is structural. It does not depend on benevolence. It does not depend on policy stability. It depends on whether keys are controlled and secured. Responsibility increases. Intervention decreases. That tradeoff is the shift.

Bitcoin also separates money from identity at the base layer. Traditional systems require identity to function. Bitcoin requires keys. Identity can still be attached through use, exchanges, or regulation, but it is not enforced by the protocol itself. This separation matters because it removes one of the most common friction points women encounter. Name changes. Documentation gaps. Transitional periods. These do not affect the protocol's ability to function.

This does not make Bitcoin invisible or anonymous. It does make it structurally indifferent to personal history. That indifference is not moral. It is architectural. For women whose lives include disruption, that distinction matters.

The shift also reframes surveillance. Bitcoin's ledger is public, but it does not operate like institutional monitoring. There is no central authority watching behavior and intervening in real time. Transactions are validated by the network, not approved by an overseer. Surveillance exists, but it is passive rather than active. No account is frozen mid-transaction. No transfer is delayed pending review.

Finality is another structural shift. Bitcoin transactions cannot be reversed by appeal or policy. This is often framed as a weakness. It is a risk. It is also a boundary. Finality removes the authority to reinterpret intent after the fact. There is no institution that can decide a transaction should not have happened.

For women, this does not mean recklessness. It means clarity. Responsibility is front-loaded rather than deferred. Decisions must be made deliberately. The system does not offer correction. It also does not offer punishment through reversal. That balance is unfamiliar in a world of constant oversight.

Bitcoin's global design also shifts how borders function. Traditional money systems embed jurisdiction into the transaction itself. Crossing borders triggers friction. Bitcoin does not recognize borders at the protocol level. The network treats all transactions the same regardless of geography. Legal consequences still exist, but enforcement is external to the system.

The fixed supply of Bitcoin represents another structural shift. The rules governing issuance are not adjustable by policy decision. This is often debated economically. Structurally, it means no authority can change supply in response to pressure. The system resists discretionary expansion.

Change itself functions differently in Bitcoin. Protocol updates require broad agreement and are slow to implement. This limits unilateral intervention. Traditional systems change through policy announcements and regulatory updates. Bitcoin changes through consensus and visibility. This does not make it static. It makes power harder to exercise without resistance.

Bitcoin's lack of convenience is part of this shift. It does not smooth mistakes. It does not provide customer service. It does not simplify responsibility. This is not a virtue. It is a design consequence. Convenience in traditional systems is paired with control. Bitcoin removes control and with it much of the convenience.

For women used to systems that promise ease while extracting compliance, this tradeoff is revealing. Bitcoin does not pretend to protect. It does not manage users. It does not intervene on their behalf. What it offers instead is a system that steps back once rules are set. This difference clarifies what Bitcoin is and is not. It is not a replacement for all money. It is not suitable for every use. It introduces new risks. It demands learning. The shift is not adoption. It is literacy.

Literacy restores choice. Women do not need to believe in Bitcoin to benefit from understanding it. They do not need to hold it to recognize what it reveals about money. Bitcoin functions as a reference point. It shows which features of traditional money are design choices rather than necessities.

Understanding Bitcoin also reframes traditional systems. Once a permissionless system is understood, permissioned systems become visible as such. Surveillance becomes legible. Reversibility becomes a form of control. Delays become mechanisms rather than inconveniences. This clarity carries back into everyday financial decisions.

The shift also restores optionality. Optionality is not escape. It is leverage. When women understand that different monetary systems operate under different rules, dependence on any single system loosens. That does not remove constraint. It reduces fragility.

This chapter does not argue that Bitcoin is safe, easy, or inevitable. It argues that Bitcoin is structurally different. That difference lies in where authority sits, how intervention works, and which rules can be changed.

The shift is seeing Bitcoin as money without managers. Money without customer service. Money without discretion. That absence is not comforting. It is clarifying.

For women navigating systems that increasingly watch, classify, and intervene, clarity matters. Bitcoin does not remove pressure. It changes the shape of it. It replaces oversight with responsibility. It replaces permission with possession. It replaces reversibility with finality.
Those replacements are not universally good. They are different. Understanding that difference is the shift.

Bitcoin does not promise freedom. It exposes structure. When women see that structure clearly, they regain the ability to decide where they want protection, where they accept risk, and where they refuse intervention. That decision belongs to them.

~5
Sending Money Across Borders

The Constraint

Sending money across borders is often described as a technical problem. Exchange rates. Fees. Delays. Compliance. These factors are treated as logistical details rather than structural barriers. In reality, cross-border money movement exposes some of the deepest assumptions embedded in modern financial systems. It reveals who money is designed to serve, who is expected to wait, and who absorbs the cost of friction.

Traditional financial systems treat borders as control points. Money does not simply move from one person to another. It passes through layers of intermediaries, each operating under different rules, jurisdictions, and incentives. Banks, correspondent banks, clearing networks, and regulators all sit between sender and recipient. Every layer adds time, cost, and scrutiny. For institutions, this complexity is manageable. For individuals, it is often destabilizing.

Remittances make this imbalance visible. Around the world, women send money across borders to support families, children, parents, and communities. These transfers are not speculative. They are not optional. They are lifelines. Yet the systems that process them treat these transactions as inherently suspicious. Frequency, destination, and urgency are read as risk signals rather than necessity.

Financial institutions are optimized for large, predictable flows. Corporate transfers. Government payments. Trade settlement. These transactions fit institutional expectations. Small, frequent remittances do not. They appear irregular. They cross jurisdictions. They often involve recipients without formal banking access. What is normal in lived experience becomes abnormal in system logic.

For women, this mismatch carries cost. Fees are higher for small transfers. Exchange rates are worse. Delays are longer. Documentation requirements are heavier. Each transaction requires effort and attention. Sending money becomes labor. That labor is invisible to the system but cumulative in real life.

Cross-border systems also amplify surveillance. Transfers are flagged based on destination, frequency, and amount. Certain regions are treated as higher risk by default. Women supporting family in those regions encounter repeated scrutiny regardless of intent. The system does not ask why support is needed. It tracks that support exists.

Delays are normalized in cross-border transfers. Funds can take days or weeks to arrive. Intermediaries perform checks at each step. Reviews compound. For women sending money to cover immediate needs, these delays are not neutral. Rent, food, medical care, and school fees do not pause while compliance processes unfold.

When transfers are delayed or blocked, responsibility is pushed back onto the sender. She must explain.

Resubmit documents. Verify identity. Justify urgency. The burden of proof rests with the individual, not the institution. This burden is often repeated with each transaction, even when patterns are consistent and benign.

Borders also create asymmetry in power. The sender operates under one regulatory regime. The recipient under another. Neither has control over the intermediaries between them. When something goes wrong, accountability dissolves across jurisdictions. Customer service becomes fragmented. No single institution owns the problem.

For women, this fragmentation increases vulnerability. She may be supporting dependents who have no alternatives. She may be navigating time zones, language barriers, and inconsistent communication. The system assumes stability and patience. Her reality often requires speed and certainty.

Cross-border controls also intersect with migration and displacement. Women who move between countries, whether temporarily or permanently, often maintain financial ties in multiple places. Accounts are flagged for cross-border activity. Transfers between personal accounts can trigger review. The system treats mobility as anomaly rather than norm.

Identity verification intensifies at borders. Names must match records exactly. Addresses must be current. Documentation must align across systems that do not share standards. Women who have changed names, relocated, or exited relationships face repeated friction. Each discrepancy delays access.

Exchange rates introduce another layer of loss. Conversion is rarely neutral. Fees are embedded. Rates are adjusted quietly. The sender often bears the cost without visibility. Over time, these losses compound. Women with limited margin absorb them silently because the alternative is not sending money at all.

Remittance systems also reinforce dependence on intermediaries. Informal networks emerge to bypass cost and delay. These networks carry their own risks. Women are forced to choose between formal systems that are slow and expensive and informal systems that are faster but less secure. The system does not offer a neutral option.

Governments frame cross-border controls as necessary for security and compliance. These goals are real. Their impact is not evenly distributed. Large institutions adapt easily. Individuals adapt through self-limitation. Women send less than needed. They delay support. They ration care across borders.

This self-limitation is a rational response to friction. It is also a constraint. The system shapes behavior not through prohibition, but through exhaustion. Each transfer requires calculation. Is the fee worth it. Will it be delayed. Will it be flagged. Over time, these questions narrow action.

Cross-border money movement also exposes how deeply money is tied to jurisdictional power.

Traditional systems assume that money should be governed by states first and individuals second. Borders are enforcement mechanisms embedded into financial infrastructure. Women experience this not as policy, but as restriction on care.

The constraint is not that remittances are impossible. It is that they are costly, slow, and uncertain by design. Systems that serve institutional stability treat personal obligation as external. Care is not prioritized. Urgency is not optimized for. Predictability matters more than consequence.

For women everywhere, sending money across borders reveals a truth that is often hidden in domestic transactions. Money is not simply a tool. It is a governed process. It moves according to rules that privilege institutions, states, and large flows over individual need.

This constraint is not theoretical. It shapes daily decisions. How much to send. When to send. Whether to send at all. It determines who absorbs risk and who is insulated from it. Women absorb the risk because the need does not disappear.

Understanding this constraint is essential. Before any alternative can be evaluated, it must be clear why cross-border money movement is so fraught in the first place. Remittances expose the limits of traditional systems. They show where money serves power rather than people. For women supporting others across borders, that exposure is constant. The system does not adapt to care. It expects care to adapt to it.

The Shift

The shift begins when women stop treating cross-border transfers as exceptions and start treating them as signals. Remittances expose how financial systems behave when jurisdiction, surveillance, and urgency collide. Once cross-border money is understood as a stress test rather than a nuisance, the system becomes legible. Confusion gives way to orientation. Expectations narrow. Surprise loses its power.

Traditional systems are not confused by remittances. They are responding exactly as designed. Borders activate oversight. Frequency activates review. Urgency activates delay. The shift is recognizing that these reactions are structural, not personal. When women stop interpreting friction as failure, clarity replaces self-blame.

This clarity changes how responsibility is held. When a transfer is delayed or blocked, the instinct is often to comply harder. Provide more documents. Explain again. Wait longer. The shift is recognizing that compliance is already assumed and that what is being enforced is control, not correctness. That recognition alters posture and planning.

The shift also reframes care. Remittances are acts of obligation, not transactions of convenience. Traditional systems do not recognize care as a legitimate financial category. They recognize risk, jurisdiction, and compliance. Seeing this mismatch allows women to stop expecting institutions to honor what they were not built to see.

This reframing restores agency. Instead of asking why urgency is not accommodated, women can ask where urgency must be protected. That question redirects effort from fixing institutions to structuring support. Agency does not require reform. It requires literacy.

The shift sharpens understanding of timing. Cross-border systems extract time deliberately through layered checks and handoffs. Delays are not incidental. They slow movement and increase oversight. When delay is recognized as a control mechanism rather than a logistical inconvenience, patience stops being the default response. Planning replaces waiting.

The shift also reframes cost. Fees and exchange losses are often treated as unfortunate side effects. In reality, they signal priority. Systems optimized for large flows penalize small ones. Women sending modest amounts absorb disproportionate loss. Recognizing this reframes cost as a structural tax on care rather than personal inefficiency.

This understanding changes how trust is allocated. Trust is often extended because institutions are regulated and familiar. The shift is recognizing that regulation aligns institutions with states, not individual need. Trust becomes conditional and selective, placed where incentives align rather than where branding reassures.

Boundaries follow naturally. Women stop expecting a single channel to serve every function. Some systems are used for integration and visibility. Others for speed and resilience. This segmentation is not avoidance. It is alignment. Different tools for different pressures.

The shift separates legitimacy from permission across borders. Traditional systems equate approved transfers with legitimate support and blocked transfers with suspicion. This framing internalizes institutional judgment. Women begin questioning whether support is appropriate rather than whether the system is constrained.

Recognizing that permission reflects policy, not morality, restores clarity. Supporting family across borders is not invalid because a system delays it. A blocked transfer does not negate obligation. Detaching legitimacy from approval allows action based on reality rather than fear of interpretation.

This distinction matters most in urgent situations. Medical needs. Housing crises. Sudden loss of income. In these moments, systems designed for review collide with lives that require response. The shift is recognizing that urgency does not register as urgency to institutions and planning must account for that gap.

The shift also reframes identity friction. Cross-border transfers intensify verification across systems that do not share standards. Women whose lives include movement encounter this repeatedly. Recognizing that mismatch is structural, not suspicious, prevents internalization of blame and reduces exhaustion.

This awareness encourages proactive organization rather than reactive explanation. Women stop scrambling to justify themselves after access is delayed. They anticipate where friction is likely and plan accordingly. Anticipation does not remove constraint, but it reduces its shock.

Visibility is also reframed. Cross-border transfers increase exposure by default as intermediaries touch each step. Records persist across jurisdictions. Women learn to distinguish necessary visibility from excessive exposure. Not every act of care requires maximum transparency. This is not secrecy. It is proportionality. Visibility should match function and consequence. When women stop assuming exposure is inherently protective, they regain discretion without abandoning responsibility.

Understanding remittances clarifies why alternative systems exist. They do not arise because borders should not exist. They arise because existing systems treat cross-border care as a problem to be managed rather than a reality to be supported. The shift is not believing alternatives are perfect. It is understanding the constraint they respond to.

This removes pressure to conform. Women do not need identities or ideologies to recognize structural differences. Different tools can serve different purposes. Integration where stability matters. Separation where responsiveness matters. Flexibility replaces loyalty.

The shift restores optionality. Optionality is not escape from borders. It is leverage within them. When women understand how money behaves across jurisdictions, dependence on any single channel loosens. That does not remove friction, but it reduces fragility. In tightening environments, this optionality becomes essential. Cross-border controls expand quietly. Thresholds lower. Reviews widen. Women who understand this are less disoriented by sudden restriction. They are not immune to constraint, but they are oriented.

The shift does not promise smooth transfers or lower fees. No system guarantees that. What it offers is clarity about tradeoffs. Formal channels provide legitimacy and integration in exchange for delay, cost, and scrutiny. Once that exchange is visible, women can decide deliberately when it is acceptable and when it is not.

This chapter does not argue for abandoning formal remittance systems. It argues for removing the assumption that they are neutral or sufficient. They are institutional tools with specific incentives. Seeing them clearly allows engagement without illusion.

The shift is practical, not ideological. It restores agency by aligning expectations with reality. When cross-border money is understood as governed infrastructure rather than simple transfer, dependence loosens and planning improves. Remittances reveal where money serves states first and people second. The shift is refusing to let that hierarchy remain invisible. When women see it clearly, they regain the ability to choose how they support others, how much friction they accept, and where they seek resilience.

~6
Inflation, Devaluation, and Quiet Theft

The Constraint

Inflation is often described as an abstract economic force. Percentages rise. Charts move. Experts debate causes and remedies. For most people, inflation is presented as a background condition, something to be managed by policy rather than experienced personally. For women who plan carefully and save deliberately, inflation is not abstract. It is a slow erosion that reshapes choices without announcing itself.

Devaluation operates quietly. Money does not disappear overnight. It loses capacity gradually. What once covered a month now covers three weeks. What once provided security now feels thin. This gradualism is important. Sudden loss provokes response. Slow loss produces adaptation. Women adjust budgets. They delay purchases. They reduce margins. Over time, adaptation masks harm.

Traditional financial systems normalize this process. Inflation is framed as necessary. Moderate devaluation is described as healthy. Savers are encouraged to accept loss as the cost of participation. The system assumes that individuals can offset erosion through growth, investment, or increased income. This assumption does not hold evenly.

Women are more likely to save defensively. Savings are held for stability rather than speculation. Emergency buffers. Caregiving gaps. Interruptions in income. These savings are not excess. They are protection against volatility. When inflation erodes these buffers, women lose resilience before they lose purchasing power.

This erosion is rarely recognized as theft because it lacks an actor. No single institution removes value directly. Loss is distributed across time and policy. Responsibility dissolves. The effect, however, is real. Money stored faithfully buys less later. Planning becomes less reliable. Discipline is punished quietly.

Inflation also undermines predictability. Long-term planning depends on stable reference points. Rent. Food. Utilities. Education. Care. When prices rise unevenly, forecasts break down. Women who plan meticulously find their assumptions invalidated not by error, but by drift. The ground moves beneath careful steps.

This instability is compounded by wage dynamics. Income does not rise uniformly with prices. Women's earnings are more likely to be interrupted or constrained. Caregiving responsibilities. Part-time work. Sectoral concentration. When costs rise faster than income, savings are used to bridge gaps. Those savings then erode further through inflation. A feedback loop forms.

Traditional advice often frames inflation as manageable through investment. Put money to work. Seek returns.

Accept risk. This advice assumes access, confidence, and margin. Many women do not have surplus capital to expose to volatility. Savings held for emergencies cannot be risked casually. The system treats caution as inefficiency.

Inflation also interacts with debt asymmetrically. Borrowers may benefit as debts are repaid with devalued currency. Savers lose as stored value declines. Women are often encouraged to carry debt strategically while maintaining minimal savings. This structure increases dependence on credit and reduces autonomy. When inflation rises, the cost of necessities increases before debt burdens adjust.

Quiet theft also operates through policy opacity. Monetary decisions are complex and distant. Central banks adjust rates. Governments expand supply. The rationale is technical. The impact is personal. Women experience consequences without visibility into cause. This distance reduces accountability and weakens resistance.

Devaluation is also uneven across goods. Essentials rise faster than discretionary items. Food, housing, healthcare, and education absorb more of women's budgets. Inflation metrics may understate lived experience. Official numbers smooth reality. Women feel pressure long before it is acknowledged.

Savings held in traditional accounts are especially vulnerable. Interest rates often lag inflation. Money grows nominally while shrinking in real terms. This creates the illusion of stability. Statements show balances intact. Value is slipping underneath. The system rewards patience with loss.

This erosion affects behavior. Women become reluctant to save. Why set money aside if it will buy less later. Consumption accelerates not out of irresponsibility, but out of rational response. If holding money guarantees loss, spending sooner feels safer. The system interprets this as lack of discipline rather than adaptation.

Inflation also redistributes power. Those with assets that appreciate benefit. Those holding cash lose. Women are less likely to hold appreciating assets due to access, risk tolerance, or life stage. The system amplifies existing inequalities while presenting outcomes as neutral.

Devaluation also complicates care. Money set aside for future needs loses adequacy. Education funds fall short. Medical reserves thin. Retirement planning becomes unstable. Women must revise plans repeatedly, often without additional income to compensate. Planning becomes labor-intensive and emotionally taxing.
The quiet nature of inflation also suppresses collective response. Because loss is gradual and individualized, it is experienced privately. Each woman adjusts alone. There is no single moment of failure to rally around. The system benefits from this dispersion.

Inflation is also framed as temporary. Spikes are expected to settle. Normalization is promised. In the meantime, women absorb loss. When normalization arrives, if it does, lost purchasing power is not restored. The baseline has shifted. What was lost remains lost.

This dynamic trains acceptance. Women learn to expect erosion. They build plans that assume decline. Ambition narrows. Safety margins shrink. The system does not need to confiscate savings directly. It allows time to do the work. Quiet theft is effective because it avoids confrontation. There is no freeze to contest. No denial to appeal. Loss arrives without paperwork. It is normalized through language and expertise. Women are told this is how money works.

The constraint is not ignorance. Many women understand inflation intellectually. The constraint is lack of control. Monetary policy operates beyond individual reach. Savings held responsibly are exposed by default. Opting out is difficult without taking on new risks.

Understanding this constraint matters. Before any discussion of alternatives or protections, the nature of the loss must be named. Inflation and devaluation are not merely economic conditions. They are mechanisms that transfer value from savers to systems that prioritize flexibility and growth.

For women who plan carefully and save deliberately, this transfer feels personal because it is. The system rewards movement, leverage, and exposure. It penalizes patience, caution, and care. Recognizing this does not solve the problem. It clarifies why the problem persists. Quiet theft is not accidental. It is tolerated because it is diffuse. It operates through policy, not force. It reshapes behavior slowly. For women navigating tight margins and long horizons, its impact is cumulative and destabilizing.

The Shift

The shift begins when women stop treating inflation as a temporary disturbance and start treating it as infrastructure. Inflation is not a mood of the economy. It is a mechanism embedded in how modern money is managed. It performs specific functions reliably and fails predictably outside those functions. Once inflation is understood this way, the relationship changes. Confusion gives way to orientation. Moral judgment recedes. Expectations narrow.

Modern monetary systems are designed for flexibility at the institutional level. Governments require the ability to respond to crises, manage debt, and stimulate activity. Inflation provides that flexibility by redistributing value quietly across time. This feature is not accidental. It is the reason inflation is tolerated and normalized. The shift is recognizing that this same flexibility makes inflation hostile to people who rely on savings for stability rather than leverage.

For women, this recognition restores agency. Inflation often feels destabilizing because it violates an unspoken promise. Save carefully. Delay gratification. Plan ahead. When savings lose value anyway, it feels like a personal miscalculation. When inflation is understood as routine system behavior, the emotional impact changes. The system is not punishing caution. It is prioritizing institutional maneuverability. Clarity replaces self-blame.

This shift also reframes discipline. Discipline is often treated as moral virtue in personal finance. Spend less. Save more. Be patient. Inflation quietly undermines this narrative. Money held responsibly loses value while risk-taking is rewarded. Recognizing this disconnect allows women to stop interpreting erosion as failure and start interpreting it as exposure to policy.

Trust is also reexamined. Trust is often placed in systems because they are longstanding and familiar. National currencies feel safe because they persist. The shift is recognizing that persistence does not imply protection of purchasing power. Monetary systems are aligned with governments and markets, not with individual savings goals. Knowing this allows trust to become conditional rather than assumed.

Boundaries follow naturally. Money held for immediate needs must behave differently than money held for long-term security. Treating all savings the same concentrates risk. Inflation exposes that concentration by eroding everything slowly and simultaneously. The shift is recognizing that segmentation is not complexity for its own sake. It is protection against uniform loss.

Control also becomes clearer through this lens. Control is not the ability to earn money. It is the ability to preserve its usefulness over time. Inflation removes that control gradually. No approval is denied. No access is frozen. Value simply thins. Recognizing this clarifies where control actually resides and where it does not.

This clarity matters more as systems normalize persistent inflation. Temporary explanations become permanent conditions. Price stability is redefined. Targets shift. Women who understand this are less likely to plan as if restoration is guaranteed. They recognize adjustment as the baseline rather than the exception. Orientation replaces hope.

Risk also looks different once inflation is named accurately. Cash savings are often framed as low risk. In reality, they carry guaranteed loss under sustained inflation. Investment is framed as risk, even when it may preserve value better over time. The shift is not dismissing risk. It is understanding which risks are visible and which are obscured. Women already understand hidden risk in other areas. Caregiving without backup. Employment without security. Health without margin. Applying the same awareness to money aligns with lived experience. It is not radical. It is consistent.

Another aspect of the shift is emotional. Inflation creates quiet anxiety. The numbers do not scream. They whisper. Budgets tighten incrementally. Plans are revised repeatedly. Women often absorb this tension alone. Naming inflation as structural rather than personal releases that pressure. Emotional energy can be redirected toward strategy rather than endurance.

The shift also separates legitimacy from reward. Traditional narratives imply that good behavior is eventually rewarded. Inflation breaks this link. Saving carefully does not guarantee preservation. Recognizing this allows women to stop waiting for fairness from systems that are not designed to provide it.

This distinction matters most over long horizons. Retirement planning. Education funds. Care reserves. Inflation compounds quietly across decades. Women who plan for long timelines are disproportionately affected. Seeing inflation clearly allows planning to account for erosion rather than be surprised by it.

Privacy is reframed as well. Inflation encourages movement of money into systems that promise growth but require visibility. Assets, accounts, and instruments increase exposure to market volatility and oversight. Women learn to weigh preservation against exposure. Not every attempt to preserve value needs maximum participation.

Privacy here is not concealment. It is proportionality. Decisions about where and how money is stored should reflect purpose and risk tolerance, not default assumptions about safety. When women stop assuming that nominal stability equals real stability, discretion returns.

Understanding inflation also clarifies why alternatives are sought. Alternatives do not arise because people misunderstand economics. They arise because existing systems accept slow loss as normal. The shift is not believing alternatives are perfect. It is recognizing the constraint they respond to. This removes pressure to conform to a single financial identity. Women do not need to become aggressive investors or passive savers. Different tools can serve different roles. Stability for daily life. Resilience for long horizons. Flexibility replaces doctrine.

The shift also encourages strategic patience. Inflation trains urgency through fear of loss. Spend now before value erodes. Chasing returns can become reactive. Seeing inflation clearly allows women to choose pacing intentionally rather than being driven by anxiety.
In tightening environments, this shift becomes essential. Inflation is often paired with increased control. Higher rates. Tighter credit. Reduced public support. Women who understand inflation as infrastructure are less likely to be disoriented by these shifts. They are not immune to pressure, but they are oriented.

The shift does not promise preservation. No monetary system can guarantee that. What it offers is clarity about tradeoffs. Fiat money offers liquidity and state backing in exchange for exposure to devaluation. Once that exchange is visible, women can decide deliberately how much exposure is acceptable.

This chapter does not argue for rejecting national currencies. It argues for removing the assumption that they are neutral stores of value. They are policy instruments with specific functions. Seeing them clearly allows engagement without illusion.

The shift is practical, not ideological. It restores agency by aligning expectations with reality. When inflation is understood as structural rather than incidental, planning becomes grounded and resentment fades. Quiet theft relies on confusion and patience. The shift interrupts both. When women see slow loss clearly, they regain the ability to choose how they save, where they absorb risk, and what they refuse to normalize.

~7
Ownership Without Permission

The Constraint

Ownership is often treated as a legal concept. Titles, accounts, deeds, and registrations define who owns what. Control is assumed to follow automatically. If something is in your name, it is yours. In practice, ownership and control are not the same. For women navigating modern financial systems, this gap is one of the most consequential constraints they face.

Traditional systems define ownership through permission. Access is granted, maintained, and revoked by institutions. Accounts exist because a bank allows them to exist. Assets are usable because platforms authorize transactions. Legal recognition is mediated through layers of compliance, policy, and oversight. Ownership is conditional, even when it appears absolute on paper.

This conditionality is usually invisible until it matters. Day to day, systems function smoothly. Transactions clear. Balances update. The illusion of control holds. When pressure appears, the distinction emerges. Accounts are frozen. Transfers are delayed. Assets become temporarily unusable. Ownership remains in name while control disappears in practice.

For women, this distinction is not theoretical. Control is most likely to be interrupted precisely when autonomy is most needed. Leaving unsafe situations. Responding to emergencies. Supporting others discreetly. Acting quickly. Traditional systems are optimized for predictability, not urgency. When urgency conflicts with compliance, compliance wins.

The constraint deepens because institutional permission is asymmetric. Institutions can revoke access instantly. Individuals must appeal slowly. There is no equivalent power on the other side. Ownership does not grant symmetry. It grants exposure.

Legal ownership also assumes stable identity and circumstance. Names match records. Addresses remain current. Documentation is available. Women's lives often violate these assumptions. Name changes. Relocation. Caregiving interruptions. Shared financial arrangements. Each deviation increases friction. Control becomes contingent on administrative alignment rather than need.

Platforms reinforce this dynamic. Increasingly, access to assets is mediated by digital interfaces governed by terms of service rather than law alone. Accounts can be suspended. Features restricted. Transactions reversed. Decisions are automated. Explanations are limited. Ownership exists within a framework of ongoing permission.

This matters because platforms define the boundary of use. You may own an asset in theory but be unable to move, sell, or transfer it without platform approval. The system recognizes ownership only insofar as it aligns with policy. Control is subordinated to rule enforcement.

The language of protection obscures this reality. Restrictions are framed as safeguards. Freezes are described as temporary. Reviews are called routine. These descriptions minimize the impact of lost control. For women managing tight margins and complex obligations, temporary loss can have permanent consequences.

Ownership without control also creates psychological vulnerability. Women plan as if assets are accessible. They assume savings can be mobilized when needed. When access is interrupted, shock compounds stress. The system's predictability disappears at the moment it is most relied upon.

This unpredictability encourages self-limiting behavior. Women avoid moving assets. They delay decisions. They maintain balances below thresholds. They accept dependence rather than risk interruption. These adaptations are rational responses to uncertainty. They also narrow autonomy over time.

Speculation is often framed as the primary financial risk women should avoid. Market volatility. Price swings. Loss of value. This framing obscures a more immediate threat. An asset that fluctuates but remains usable may preserve autonomy better than a stable asset that becomes inaccessible under pressure. For many women, control matters more than price movement.

Joint ownership complicates this further. Many women share accounts, property, or credit lines. Legal structures rarely reflect power dynamics inside these arrangements. When conflict arises, formal ownership rules determine access. Contribution, caregiving labor, and dependency are invisible. Control follows paperwork, not reality.

The constraint is reinforced by time. Appeals take days or weeks. Documentation must be gathered. Reviews proceed sequentially. During this period, ownership exists without usability. Women are expected to wait calmly while obligations continue. The system treats time as neutral. It is not.

Control also erodes through policy drift. Rules change. Thresholds adjust. Reporting expands. What was permitted yesterday becomes restricted today. Ownership remains nominally intact while usable control shrinks. Women absorb this erosion gradually, often without a clear point of resistance. This drift trains acceptance. Women learn to expect friction. They plan around limitations. They internalize constraints as normal. Over time, autonomy is reduced not by explicit prohibition but by cumulative adjustment.

Ownership without permission is rare in traditional systems. Even physical assets are increasingly mediated by regulation, reporting, and platform integration. Money, property, and digital assets are embedded in networks of oversight. Control is shared with institutions by default.

This structure benefits stability at scale. It allows coordination, enforcement, and recovery. It also creates a hierarchy of access. Those aligned with institutional norms experience fewer interruptions. Those operating at the margins encounter more scrutiny. Women are overrepresented at those margins due to caregiving roles, interrupted careers, and relational financial responsibilities.

The constraint is not lack of sophistication. Women often understand these systems intuitively. The constraint is lack of exit. Fully opting out of permissioned ownership is difficult without accepting new risks. Most systems require tradeoffs that are not evenly distributed. Understanding this constraint matters. Before discussing alternatives or shifts, it must be clear why ownership feels fragile even when assets are substantial. Control, not value, determines autonomy. Without control, ownership is symbolic.

This chapter names that constraint. Ownership mediated by permission prioritizes institutional order over individual autonomy. It rewards compliance and predictability. It penalizes urgency and discretion. For women navigating complex lives, this structure shapes every financial decision.

Ownership without control is incomplete ownership. Recognizing that gap is the first step toward understanding why control matters more than speculation.

The Shift

The shift begins when women stop treating ownership as a legal status and start treating it as a relationship to control. Ownership on paper feels reassuring because it is legible and recognized. Control is quieter. It appears only when something needs to be moved, accessed, or protected. Once this distinction is understood, the relationship to assets changes. Expectations narrow. Surprise loses its power.

Traditional systems are designed to recognize ownership without guaranteeing usability. Titles, accounts, and registrations establish legitimacy, not access under pressure. The shift is recognizing that legitimacy and control are separate functions. Institutions are built to certify ownership. They are built to manage access according to policy. They are not built to prioritize individual timing or discretion.

For women, this recognition restores agency. Many access interruptions feel destabilizing because they violate an assumption of reliability. If something is owned, it should be usable. When that assumption breaks, it feels like failure. When the interruption is understood as routine enforcement behavior, the emotional impact changes. The system is not reacting to her needs. It is executing rules. Clarity replaces self-blame.

This shift also reframes trust. Trust is often extended to institutions because ownership is formalized through them. A long-standing account or platform relationship feels secure because it is familiar. Familiarity is not alignment. Institutions are aligned with compliance, liability management, and systemic stability. They are not aligned with personal autonomy. Recognizing this allows women to place trust selectively rather than globally.

Boundaries follow naturally. Permissioned systems are reliable for predictable, visible activity that can tolerate delay and review. They are unreliable when speed, discretion, or independence matter. Seeing this distinction allows women to plan around known fault lines instead of assuming universal functionality. Planning replaces hope. Hope is fragile in systems that do not reciprocate it.

The shift sharpens the meaning of control. Control is not whose name appears on an asset. It is who can interrupt access, reverse movement, or demand explanation. Traditional systems concentrate that authority within institutions. Once this is acknowledged, women can decide intentionally which resources must remain immediately usable and which can tolerate oversight.

This distinction becomes more important as systems tighten. Permission rarely loosens. It expands. Automation replaces discretion. Reviews become opaque. Appeals slow or disappear. Women who understand this are less likely to interpret new restrictions as anomalies. They recognize enforcement patterns and adjust expectations accordingly. Orientation replaces shock.

Risk also looks different through this lens. Volatility is visible and dramatic. Control loss is quiet and procedural. An asset that fluctuates in value but remains usable preserves options. An asset that is stable but inaccessible removes them. The shift is not dismissing price risk. It is recognizing access risk as equally consequential.

Another aspect of the shift is emotional. Ownership failures often trigger shame. Why was access removed. Why did this happen now. Reframing these events as predictable outcomes of system design removes that burden. Emotional energy can be redirected toward response rather than justification. Calm returns when meaning is clarified.

The shift also separates legitimacy from permission. Traditional systems equate approval with correctness. If access is granted, use is legitimate. If access is blocked, suspicion follows. This framing internalizes institutional judgment. Women begin questioning their own decisions rather than the system's constraints. Recognizing that permission reflects policy, not morality, restores clarity.

This distinction matters most in moments of urgency. A blocked transfer does not invalidate the need it was meant to meet. A frozen account does not imply wrongdoing. When legitimacy is detached from approval, women can act based on reality rather than fear of interpretation.

Privacy is reframed as well. Permissioned ownership requires visibility. Visibility supports oversight and enforcement. That visibility does not serve personal safety equally. Women already practice selective disclosure in professional, medical, and social contexts. Extending that practice to ownership is not secrecy. It is proportionality.

Privacy here is about scale and permanence. Not every asset needs continuous monitoring. Not every movement needs institutional exposure. When women stop assuming that full transparency is inherently protective, discretion returns without abandoning responsibility.

Understanding ownership constraints also clarifies why alternative forms of holding and control exist. Alternatives do not arise because institutions are broken. They arise because institutions are effective at serving priorities that do not include individual autonomy under pressure. The shift is not believing alternatives are perfect. It is understanding the constraint they respond to.

This removes pressure to conform to a single ownership model. Women do not need to adopt identities or ideologies to use multiple systems. Different tools can serve different purposes. Certification where legitimacy matters. Control where autonomy matters. Flexibility replaces loyalty. Loyalty is rarely rewarded by systems designed for scale rather than reciprocity.

The shift encourages strategic engagement. Women can continue using permissioned systems for predictable functions while ensuring that critical resources are not entirely dependent on institutional approval. This is not avoidance. It is segmentation. Segmentation reduces exposure without demanding withdrawal.

In high-control environments, this shift becomes essential. Authority rarely announces itself directly. It expands through systems that already manage access and compliance. Ownership becomes conditional through infrastructure rather than decree. Women who understand this are not immune to constraint, but they are less disoriented by it.

The shift does not promise unrestricted control. No system can. What it offers is clarity about tradeoffs. Permissioned ownership provides recognition and integration in exchange for conditional access. Once that exchange is visible, women can decide deliberately when it is acceptable and when it is not.

This chapter does not argue for abandoning traditional ownership frameworks. It argues for removing the assumption that ownership guarantees control. Institutions certify status. They do not guarantee usability. Seeing this clearly allows engagement without illusion.

The shift is practical, not ideological. It restores agency by aligning expectations with reality. When ownership is understood as contingent unless control is secured elsewhere, dependence loosens and planning improves. Ownership without permission is not about speculation or rebellion. It is about preserving the ability to act when it matters. When women see where control actually resides, they regain the ability to decide which risks to accept and which to refuse.

~8
Independence Without Isolation

The Constraint

Independence is often framed as separation. Self-sufficiency. Financial autonomy. Standing alone. This framing carries a quiet cost. It suggests that needing others is weakness and that reliance erodes agency. For women, whose lives are often shaped by care, coordination, and shared responsibility, this framing creates a false choice between independence and connection.

Traditional financial systems reinforce this false choice. They are built around individual accounts, individual credit, and individual liability. Support that flows between people is treated as exception rather than norm. Shared responsibility is flattened into transactions. Care becomes an anomaly. The system recognizes individuals. It does not recognize networks.

This mismatch creates constraint. Women often manage resources relationally. Supporting children, elders, partners, friends, and community members. Money moves to meet need rather than to maximize return. Traditional systems interpret this movement through a lens of risk and dependency. Shared flows are flagged. Informal support is scrutinized. Independence is rewarded. Interdependence is questioned.

The constraint deepens because institutional independence is defined narrowly. Independence means having your own account. Your own credit history. Your own documentation. It does not mean having control over resources within a network. Women may be financially independent on paper while being deeply constrained in practice by shared obligations that the system does not accommodate.

Mutual aid exposes this tension clearly. Informal support networks arise where formal systems fall short. Childcare swaps. Emergency funds. Community support during crisis. These arrangements rely on trust, speed, and discretion. Traditional financial systems are poorly suited to support them. Transfers are slow. Visibility is high. Compliance requirements interrupt urgency.

When women move money to support others quickly, the system often intervenes. Frequency triggers review. Destinations raise questions. Accounts are flagged. The system treats coordination as anomaly. It does not recognize care as infrastructure. The result is friction precisely where resilience is being built.

Isolation is often presented as the cost of independence. Be careful. Protect yourself. Do not entangle finances. These warnings are not wrong. Financial entanglement carries risk. The problem is that the system offers isolation as the only safe alternative. It does not provide tools for shared autonomy. Women are left choosing between exposure and withdrawal.

This choice is particularly stark in moments of crisis. Illness. Job loss. Displacement. Natural disaster. Community response requires rapid pooling and redistribution of resources. Traditional systems slow this response. Approvals delay action. Platforms impose limits. Women must navigate compliance while needs escalate.

The constraint is reinforced by surveillance. Financial systems monitor behavior at the individual level. Patterns of support become data points. Networks are rendered visible without context. What is solidarity in practice becomes suspicion in policy. Women learn to self-limit support to avoid scrutiny. This self-limiting narrows community capacity. Support is rationed. Requests are delayed. Help becomes conditional on system tolerance rather than need. Over time, networks weaken. Isolation grows not because women choose it, but because coordination becomes costly.

Traditional systems also individualize risk. Each person is assessed separately. Credit scores. Account histories. Liability structures. Shared resilience is not recognized. When women contribute to community stability, the system records only individual exposure. The benefits of mutual aid are invisible. The risks are fully assigned.

This invisibility discourages collective action. Women are told to protect their own standing first. Do not jeopardize your account. Do not trigger review. Do not co-sign. These warnings reflect real risk. They also reinforce a system that cannot distinguish exploitation from solidarity.

Family support is similarly constrained. Women often manage resources across households. Supporting children in different locations. Assisting parents. Coordinating care expenses. Traditional systems treat these flows as cross-account transfers without recognizing shared purpose. The system assumes independence. Women live interdependence.

The language of responsibility compounds the problem. Be financially independent. Do not rely on others. This language ignores the reality that care is shared by necessity. Children, elders, and communities do not operate as individuals. Women are tasked with bridging this gap using tools designed for isolation. This gap produces emotional strain. Women feel torn between prudence and generosity. Between protecting themselves and showing up for others. When support creates friction, guilt follows. When isolation feels safer, loss follows. The system offers no resolution because it was not built to support relational autonomy.

Independence within traditional systems often requires invisibility of support. Women compartmentalize. They hide assistance. They move money indirectly. They delay help until it appears less risky. These adaptations preserve access but erode trust within networks. Transparency inside communities is sacrificed to opacity demanded by institutions.

This is not because independence and connection are incompatible. It is because the system treats them as such. Tools that support individual control often undermine collective resilience. Tools that support collective action trigger individual risk.

The constraint is not that women value independence too much or connection too much. It is that they are asked to choose between them. The system does not allow independence that includes others. It allows autonomy only when it is isolated.

This becomes more acute as systems tighten. Surveillance increases. Compliance expands. Informal coordination becomes harder to sustain. Women who rely on networks for resilience feel pressure to formalize or dissolve them. Formalization brings oversight. Dissolution brings isolation. The result is fragility. Individuals appear independent while networks weaken. When shock arrives, there is less capacity to respond collectively. Women absorb the impact privately. The system registers stability until it does not.

Understanding this constraint matters. Before discussing how Bitcoin might support community and mutual aid, it must be clear what traditional systems make difficult. Independence is framed as separation. Interdependence is treated as risk. Women are forced to navigate between them without adequate tools.

This chapter names that constraint. Independence without isolation is possible, but it is not supported by systems built to monitor individuals rather than enable networks. Recognizing this gap is the first step toward understanding why alternative structures are explored at all. The problem is not that women seek connection. It is that connection has been made costly by design.

The Shift

The shift begins when women stop treating independence as separation and start treating it as control within relationship. Independence does not require standing alone. It requires the ability to act without asking permission. Once this distinction is made, independence and connection stop competing. They become complementary. Orientation replaces tension. False choices lose their force.

Traditional systems frame autonomy as isolation because isolation is easier to govern. Individuals are legible. Networks are not. The shift is recognizing that this framing serves institutional simplicity, not human reality. Women do not need to abandon connection to preserve agency. They need tools that allow coordination without surrendering control.

This reframing restores agency immediately. Many women feel constrained not because they lack resources, but because helping others exposes them to scrutiny and interruption. When independence is understood as control over timing, movement, and access, generosity no longer feels reckless. Support becomes intentional rather than hidden.

The shift also reframes mutual aid. Mutual aid is often treated as informal or temporary, something that exists only in emergencies. In reality, it is continuous infrastructure. Childcare. Transportation. Food. Housing support. Emotional labor. Women already participate in these systems daily. The problem is not willingness. It is tooling.

Bitcoin enters this conversation not as a solution to care, but as a different coordination layer. It does not ask permission to move value. It does not evaluate relationships. It does not distinguish between family, friend, or stranger. This indifference is structural. It allows support to move without interpretation.

For women, this matters because it removes judgment from the act of helping. Support does not need to fit institutional categories to be allowed. Frequency does not trigger review. Urgency does not invite delay. Coordination does not become visibility by default. Control remains with the participants.

This does not eliminate responsibility. It relocates it. Responsibility shifts from compliance with external policy to care within the network. Women decide who to support, how much, and when. The system does not intervene. That absence of intervention changes the emotional texture of support. Helping does not feel like risk management. It feels like action.

The shift also reframes trust. Trust is often confused with institutional backing. A system feels trustworthy because it is regulated and familiar. Bitcoin replaces institutional trust with procedural trust. Rules are known. Enforcement is automatic. There is no discretion to appeal to and no relationship to manage. For networks, this predictability matters.

Predictability allows planning. Women can coordinate resources without worrying that support will be interrupted midstream. They can respond quickly without triggering review. They can act first and explain later within their own relationships rather than to an institution that does not share context.

Boundaries still matter. Independence without isolation does not mean unlimited sharing. It means shared autonomy. Women can support others without entangling identities or exposing unrelated resources. Control is granular. Participation is voluntary. Exit is possible. These features preserve independence while enabling connection.

The shift also changes how privacy is understood. Privacy is not secrecy from community. It is protection from unnecessary exposure. Bitcoin allows transparency within networks without broadcasting to institutions. Records exist, but they are not tied automatically to identity or hierarchy. Women can choose what to reveal and to whom. This proportional privacy strengthens trust inside communities. When support does not require concealment from helpers, coordination improves. Transparency inside the network is no longer sacrificed to satisfy external oversight. Relationships stabilize when help does not create hidden risk.

The shift reframes scale as well. Mutual aid is often imagined as small and local. Bitcoin allows coordination across distance without adding friction. Family support across borders. Community fundraising. Crisis response. Distance does not introduce new intermediaries. Networks remain intact even when geography changes.

This matters for women whose communities are dispersed. Migration, mobility, and displacement fracture traditional support structures. Bitcoin does not restore community on its own. It allows existing relationships to remain functional despite separation. Independence travels with the individual without severing ties.

The shift also separates care from charity. Charity often flows top-down. Mutual aid flows laterally. Traditional systems favor charity because it is legible and controlled. Mutual aid is relational and adaptive. Bitcoin supports lateral flows by design. There is no central allocator. Networks decide. This changes power dynamics. Support does not require institutional endorsement. Women do not need to justify need to an authority. They coordinate directly. Dignity is preserved because permission is not requested.

The shift does not deny risk. Independence without isolation still requires judgment. Trust can be misplaced. Mistakes can occur. Bitcoin does not prevent that. It removes one category of risk while leaving others intact. Women remain responsible for discernment. What changes is the absence of institutional interference.

This distinction is crucial. The goal is not safety guaranteed by systems. It is resilience built through relationships. Traditional systems promise protection but undermine coordination. Bitcoin removes protection and with it removes obstruction. The tradeoff is clear and deliberate.

In tightening environments, this shift becomes more important. Surveillance expands. Informal coordination is scrutinized. Networks are pressured to formalize or dissolve. Bitcoin allows networks to persist without becoming institutions. Independence is preserved without forcing isolation. This does not mean abandoning traditional systems. Banks still serve functions. Platforms still facilitate scale. The shift is recognizing limits. Independence without isolation requires multiple tools. Some for integration. Some for coordination. Women choose based on context, not loyalty.

The shift also restores emotional balance. Helping others no longer feels like a threat to personal stability. Independence no longer feels like withdrawal. Women can show up without sacrificing control. That alignment reduces guilt and fatigue.

This chapter does not argue that Bitcoin creates community. Community already exists. It argues that Bitcoin can support existing networks by removing friction that isolates individuals. Independence is preserved not by standing apart, but by choosing how and when to stand together.

Independence without isolation is not a slogan. It is a structural possibility. When control replaces permission and coordination replaces surveillance, women no longer have to choose between autonomy and care. That possibility is the shift.

~9
Risks, Limits, and Real Tradeoffs

The Constraint

Bitcoin is often discussed in extremes. It is either framed as a cure-all or dismissed as reckless and harmful. Both framings obscure what matters most. Bitcoin is neither a miracle nor a menace. It is a system with specific properties, costs, and limits. Understanding those limits is essential, especially for women who cannot afford abstractions that ignore consequence.

The first constraint is that Bitcoin does not eliminate risk. It redistributes it. Traditional financial systems concentrate risk within institutions and policy decisions. Bitcoin shifts risk toward individuals and networks. This shift can increase autonomy. It can also increase responsibility. Mistakes are not reversible by appeal. Errors carry weight. For women already managing complexity, this matters.

Bitcoin also does not replace the need for judgment. It does not protect against poor decisions, misplaced trust, or coercive relationships. Control over assets does not automatically translate into safety. In some situations, irreversible control can increase vulnerability. Bitcoin removes intermediaries. It does not remove power dynamics between people.

Energy use is one of the most visible and contested constraints. Bitcoin's proof-of-work system consumes electricity by design. This consumption is often framed as wasteful. The framing is incomplete. Energy use is the mechanism that secures the network. It makes rewriting history costly. It anchors digital scarcity in physical reality. Without this expenditure, the system would be easier to manipulate.

That does not make energy use irrelevant. It makes it purposeful. The question is not whether Bitcoin uses energy. It is whether that energy use is justified relative to the system it secures and the alternatives it competes with. Traditional monetary systems also consume vast amounts of energy. Data centers. Office towers. Payment networks. Military enforcement of currency regimes. Inflationary systems incentivize constant growth and consumption. These costs are diffuse and rarely counted.

Bitcoin's energy use is transparent and bounded by incentives. Miners seek the cheapest available energy. This drives use toward surplus, stranded, or renewable sources over time. Energy that would otherwise be wasted becomes valuable. The network does not require perpetual growth in energy consumption. Efficiency improvements and market pressures constrain it.

This does not mean Bitcoin is environmentally neutral. It means the comparison is often dishonest. Critiques focus narrowly on Bitcoin's energy use while ignoring the environmental costs of legacy systems that underpin global finance. Those systems externalize harm quietly. Bitcoin concentrates cost visibly. Visibility invites scrutiny. Scrutiny is not the same as condemnation.

Another constraint is volatility. Bitcoin's price fluctuates significantly. This volatility makes it unsuitable for certain uses, especially short-term stability. Women relying on predictable purchasing power must account for this. Holding Bitcoin without understanding volatility can introduce stress rather than resilience. It is not a savings account in the traditional sense.

Volatility is partly a function of Bitcoin's openness. It trades globally, continuously, without central stabilization. There is no authority smoothing price or absorbing shocks. This lack of intervention is a feature. It is also a cost. Price reflects demand and uncertainty directly. For women accustomed to systems that obscure instability until it is severe, this transparency can feel harsh.

Bitcoin also does not integrate seamlessly with existing systems. Converting between Bitcoin and traditional currencies often requires intermediaries. These points of conversion reintroduce many of the constraints Bitcoin avoids. KYC requirements. Account freezes. Platform risk. Women must navigate these edges carefully. Bitcoin does not eliminate interaction with traditional systems. It coexists with them.

Custody is another real constraint. Holding one's own keys preserves control. It also introduces risk of loss. Keys can be misplaced. Devices can fail. Backups can be compromised. For women managing caregiving, displacement, or crisis, secure custody requires planning and discipline. There is no customer service to recover access.

Delegating custody reduces this burden but reintroduces permission. Exchanges and custodial services can freeze accounts, restrict withdrawals, or fail outright. Women must weigh convenience against control. This tradeoff is not abstract. It has real consequences in moments of stress.

Bitcoin also does not solve inequality. Access to technology, education, and spare resources affects who can use it effectively. Early adopters benefit disproportionately. Knowledge gaps matter. Women already navigating systemic disadvantage must be cautious about narratives that promise equalization without addressing these realities.

The network is also not private by default. Transactions are public. While identities are not inherently attached, patterns can be analyzed. Privacy requires additional tools and understanding. Misuse can expose information unintentionally. Bitcoin reduces some forms of surveillance. It does not eliminate all visibility.

Bitcoin also operates within legal and political environments. Governments can regulate onramps and offramps. They can tax, restrict, or criminalize certain uses. Bitcoin resists direct control, but it does not exist outside power structures entirely. Women must remain aware of jurisdictional realities.

Another constraint is cultural. Bitcoin communities can be exclusionary, technical, or dismissive of lived experience. Women entering these spaces may encounter hostility or simplification. This social barrier matters. A tool that requires participation in unwelcoming environments limits accessibility.

Bitcoin also demands time and attention. Learning how it works. Securing keys. Understanding tradeoffs. This cognitive load is real. Women balancing work, care, and recovery may not have spare capacity. Any system that requires constant vigilance imposes cost.

Finally, Bitcoin does not eliminate the need for trust. It shifts trust from institutions to software, incentives, and mathematics. This trust is different, not absent. Women must decide whether this form of trust aligns with their tolerance for risk and responsibility. The constraint is not that Bitcoin is flawed. All systems are. The constraint is that Bitcoin's strengths are paired with real costs. Ignoring those costs produces harm. Overstating benefits creates false confidence. Women do not need promises. They need clarity.

This chapter names what Bitcoin does not solve. It does not guarantee safety. It does not remove risk. It does not replace community judgment. It does not absolve responsibility. It offers specific advantages in control, permission, and coordination. It also introduces volatility, complexity, and new forms of exposure.

Understanding these limits is not discouragement. It is respect. For women navigating constrained environments, honest assessment matters more than enthusiasm. Tradeoffs must be visible to be chosen. This is the constraint.

The Shift

The shift begins when women stop asking whether Bitcoin is good or bad and start asking what it does and does not do. Moral framing obscures practical judgment. Bitcoin is not an identity or a promise. It is infrastructure with specific properties. Once it is evaluated as such, tradeoffs become visible. Decision-making becomes grounded rather than reactive.

This reframing restores agency. Many discussions about Bitcoin collapse into certainty on either side. Absolute faith or absolute rejection. Neither position serves women who must manage consequence. The shift is accepting that usefulness and risk can coexist. A system can be valuable in some contexts and unsuitable in others. That recognition opens space for selective use.

Energy use is one place where this clarity matters. Bitcoin's energy consumption is real and intentional. It secures the network by making attacks costly. That cost anchors digital ownership to physical constraint. The shift is understanding energy use as part of the security model rather than an incidental byproduct. Security is never free. It is paid for somewhere.

This does not absolve Bitcoin of environmental responsibility. It contextualizes it. Energy markets reward efficiency. Bitcoin mining follows those incentives. Over time, miners migrate toward surplus, stranded, or renewable energy because it is cheapest. This trend does not eliminate impact. It shapes it. The comparison must include the environmental costs of legacy monetary systems that rely on continuous expansion, physical infrastructure, and enforcement. Those costs are diffuse and rarely attributed to money itself.

Seeing energy honestly allows proportional judgment. Bitcoin's footprint is visible and measurable. That visibility invites scrutiny and improvement. Traditional systems hide their costs behind complexity. The shift is refusing selective accounting. Women can weigh Bitcoin's energy use against the environmental reality of alternatives rather than against an imagined baseline of zero cost.

Volatility requires similar reframing. Price instability is often treated as disqualifying. In reality, volatility signals exposure to open markets without stabilization. Bitcoin does not smooth outcomes. It reveals them. This can be destabilizing for short-term needs. It can be tolerable for long-term positioning when understood deliberately. The shift is not minimizing volatility. It is planning around it. Women can distinguish between money needed for stability and assets held for optionality. Bitcoin is not suited for every purpose. Treating it as such creates harm. Treating it as one tool among many restores balance.

Custody is another area where realism matters. Self-custody preserves control and removes permission. It also transfers responsibility fully. Loss is final. Errors are permanent. This is not a flaw. It is a tradeoff. The shift is acknowledging whether that tradeoff aligns with current capacity.

For some women, full self-custody will be appropriate. For others, partial delegation may be necessary. Convenience reduces cognitive load but reintroduces institutional risk. Neither choice is pure. The shift is choosing consciously rather than defaulting.

Bitcoin's interaction with traditional systems also requires realism. Onramps and offramps remain points of vulnerability. Regulation, freezes, and platform failure still exist at these edges. Bitcoin does not dissolve the surrounding environment. It changes the core. The shift is not expecting isolation from power, but reducing dependence on it where possible. This awareness prevents overextension. Women do not need to move everything at once. Gradual engagement reduces risk. Understanding how Bitcoin behaves at boundaries matters more than ideological commitment. Caution here is not hesitation. It is calibration.

Privacy also requires active choice. Bitcoin's ledger is transparent. Privacy is not automatic. It must be learned and practiced. The shift is recognizing that Bitcoin reduces certain forms of surveillance while leaving others intact. Assuming invisibility creates exposure. Understanding visibility creates discretion.

Legal context matters as well. Bitcoin resists direct control but does not exist outside law. Jurisdictions differ. Enforcement varies. Women must remain attentive to where they live and how rules are applied. The shift is not assuming immunity. It is avoiding surprise.

Social context also matters. Bitcoin culture can be exclusionary or dismissive. This is not inherent to the technology. It affects access. Women should not feel obligated to participate in spaces that undermine their confidence or safety. Tools can be learned without adopting cultures wholesale.

Time and attention are real costs. Learning Bitcoin requires effort. Securing assets requires vigilance. These demands compete with caregiving, work, and recovery. The shift is valuing capacity honestly. Adoption that exceeds capacity creates fragility. Selective learning preserves resilience.

Bitcoin also does not eliminate inequality. Early access, technical literacy, and spare resources shape outcomes. Narratives of equalization obscure this. The shift is rejecting promises that bypass structural reality. Bitcoin can reduce certain asymmetries. It does not erase all of them. This realism allows Bitcoin to be used without myth. It can support autonomy without being burdened with expectation it cannot meet. It can coexist with other systems without replacing them. Women retain the right to step back when risk outweighs benefit.

The shift also reframes caution. Caution is often framed as fear. In reality, it is competence. Women already manage risk across domains daily. Applying that same discernment to Bitcoin reflects strength, not hesitation.

This chapter does not argue for restraint or enthusiasm. It argues for proportion. Bitcoin's strengths are specific. Its limits are real. Women do not need certainty. They need clarity.

The shift is accepting tradeoffs without defensiveness. Energy use secures the network and carries cost. Volatility exposes markets and creates instability. Control increases autonomy and responsibility. None of these cancel the others. They coexist.

When women see Bitcoin without distortion, they regain choice. They can decide where its properties align with their needs and where they do not. They can use it without relying on it. They can step away without regret. Bitcoin does not solve everything. It does not need to. The shift is letting it be what it is. Infrastructure with costs. A tool with limits. An option, not a mandate. That clarity is the shift.

~10
Choosing When and How to Use It

The Constraint

Choice is often framed as access. If a tool exists and is available, choice is assumed to follow naturally. In practice, choice requires more than availability. It requires clarity about consequence. It requires understanding tradeoffs well enough to decide deliberately rather than reactively. For women navigating constrained systems, choice without context is not empowerment. It is pressure.

Bitcoin is frequently introduced as an answer rather than an option. It is presented as a stance to adopt or reject, an identity to assume, or a future to commit to. This framing creates unnecessary friction. It forces women to decide everything at once. Believe or dismiss. Enter fully or stay out entirely. The system of presentation mirrors the rigidity many women are already trying to escape.

The constraint begins here. When Bitcoin is framed as identity, the cost of engagement rises. Identity demands alignment. It demands defense. It invites judgment. Women who approach tools pragmatically are pushed into positions they did not seek. The question shifts from usefulness to loyalty. That shift undermines agency.

Traditional financial systems already impose identity.
Account holder. Credit profile. Risk category.
Compliance subject. Bitcoin does not require identity at
the protocol level, but the surrounding discourse often
does. Communities form norms. Narratives harden.
Belonging becomes conditional. For women who value
flexibility, this environment can be alienating.

This social constraint matters because it shapes
learning. Women are less likely to explore tools that
require ideological buy-in. Curiosity narrows when
stakes feel existential. A system that demands belief
discourages experimentation. Bitcoin becomes harder
to approach not because it is complex, but because it is
framed as totalizing.

The constraint is compounded by urgency narratives.
Use it now. Everything is collapsing. Opt in or be left
behind. These narratives mirror the fear-based
messaging women already experience in financial
planning. They compress decision-making timelines.
They discourage gradual learning. They reward
impulsivity over discernment. For women managing
limited margin, compressed timelines are dangerous.
Decisions made under pressure tend to favor simplicity
over fit. Bitcoin is not simple in practice. It has edges,
responsibilities, and consequences. Treating it as an
emergency response increases risk rather than reducing
it.

Another constraint arises from comparison framing.
Bitcoin is often positioned against all traditional systems
at once. Bank replacement. Currency replacement.
Savings replacement. This framing forces Bitcoin to

carry more weight than it should. It becomes responsible for solving problems it was not designed to address. When it inevitably falls short, trust erodes.

Women are especially vulnerable to this overloading. They are often tasked with making systems work across domains. Household finance. Care coordination. Crisis response. Adding another system that promises everything but delivers selectively creates friction. The issue is not Bitcoin's limits. It is the expectation mismatch.

The constraint also appears in moral framing. Bitcoin use is sometimes portrayed as enlightened or irresponsible depending on the audience. Energy debates. Speculation narratives. Political associations. Women are asked to justify interest before understanding function. This inversion discourages practical evaluation. Morality precedes literacy.

This is familiar terrain. Women are often expected to justify financial decisions emotionally rather than structurally. Why do you need this. Is it safe. Is it appropriate. These questions obscure design and tradeoffs. Bitcoin becomes another site where intention is interrogated before utility is assessed.

The constraint deepens because learning is rarely modular. Educational resources often assume commitment. Tutorials jump from basics to evangelism. Caution is framed as doubt. Partial use is framed as misunderstanding. Women who want to learn selectively encounter resistance.

This resistance is subtle. Language implies that real understanding requires full participation. That holding some Bitcoin means holding more. That using it occasionally is missing the point. These implications pressure women to escalate involvement beyond comfort or need.

Traditional systems also reinforce this constraint indirectly. When Bitcoin is contrasted with banks as a moral alternative, women who continue using banks feel compromised. The false binary reappears. Use this or that. Independence or compliance. The reality is more layered. Women operate across systems because life requires it. The inability to hold layered positions is the constraint. Bitcoin as identity collapses nuance. It discourages segmentation. It pushes toward singular solutions in contexts that demand plural ones.

Another limitation arises from skill mismatch. Bitcoin requires technical literacy and attention. Some women will have capacity and interest. Others will not. Framing Bitcoin as identity devalues partial competence. It suggests that if one cannot manage everything, one should manage nothing.

This mirrors patterns women encounter elsewhere. Technology adoption that assumes uninterrupted focus. Financial advice that assumes linear careers. The burden of mastery is placed on individuals without regard for competing responsibilities.

The constraint is not that Bitcoin is difficult. It is that it is often presented as indivisible. All or nothing. This framing discourages adaptive use. It undermines the very autonomy Bitcoin is meant to support. Risk communication also suffers under identity framing. Limits are downplayed. Tradeoffs are minimized. Criticism is interpreted as hostility. Women seeking honest assessment encounter defensiveness. The environment becomes less informative and more performative.

This matters because women often manage downside risk. They are tasked with ensuring continuity. They absorb consequences when systems fail. They need clear boundaries more than grand narratives. A tool that cannot be discussed critically is not a tool. It is a belief system.

The final constraint is internal. When Bitcoin is framed as identity, women must decide who they are before deciding what they need. This inversion burdens decision-making. Identity should follow use, not precede it. Tools should serve life, not define it. Understanding this constraint reframes the task ahead. The question is not whether to become a Bitcoin user. It is when and how Bitcoin might be useful. Under what conditions. For which purposes. With what limits.

This chapter names the pressure that distorts choice. Bitcoin as identity narrows agency. Bitcoin as option expands it. Before discussing how to choose deliberately, it must be clear why that choice has been made artificially difficult. The constraint is not lack of access. It is excess framing. When belief replaces evaluation, women are forced into positions they do not need to occupy.

The Shift

The shift begins when women stop treating Bitcoin as a commitment and start treating it as a choice that can be made quietly, partially, and on their own terms. Choice does not require announcement. It does not require belief. It does not require permanence. Once Bitcoin is approached this way, the pressure surrounding it loosens. The tool becomes available without demanding allegiance. That availability restores agency.

This reframing changes how decision making feels. Women are no longer asked to predict the future or declare a position. They are asked to assess the present. What is needed now. What risks exist now. What constraints are active now. Bitcoin does not have to answer every question to be useful. It only has to answer one specific need at a specific moment.

Traditional systems rarely allow this kind of selective use. Accounts, credit, and platforms tend to bundle services and expectations together. Participation is total or not at all. Bitcoin behaves differently. It does not expand unless invited. It does not penalize limited use. Women can engage at the edge without being pulled inward. That property matters in environments where overcommitment creates exposure.

The shift also slows the timeline. Bitcoin does not expire if ignored. There is no requirement to act quickly. Urgency is imposed by narratives, not by the system itself. When urgency is removed, learning becomes possible. Women can observe how the system behaves without placing themselves inside it immediately. Curiosity replaces pressure. This slower pace is protective. Women often carry responsibility for continuity. Households, caregiving, and coordination do not pause for experimentation gone wrong. A tool that allows gradual familiarity respects that reality. Bitcoin can be approached as observation first, use later, or not at all. Each option remains valid.

The shift also dissolves false binaries. Women do not have to choose between Bitcoin and traditional finance, or between autonomy and participation. Systems can coexist without conflict when their roles are clear. Bitcoin can be used where permission, reversibility, or visibility are liabilities. Traditional systems can remain in place where predictability and integration are required. Segmentation becomes a strategy rather than a compromise.

This layered approach mirrors how women already operate. Formal systems for wages and taxes. Informal systems for care and support. Public roles and private coordination. Applying the same logic to money is not radical. It is consistent. Bitcoin becomes one tool among many, not a replacement for everything else.

Choosing when to use Bitcoin also requires clarity about what it does not provide. It does not offer customer service. It does not adapt to personal circumstance. It does not forgive mistakes. These limits matter. The shift is not ignoring them. It is planning around them. Women can reserve Bitcoin for contexts where these absences are acceptable or even beneficial.

This clarity restores trust in self-judgment. Women are not required to justify use or non-use to anyone. Financial tools do not confer virtue. They serve functions. If Bitcoin does not fit a current need, declining it is not failure. It is discernment exercised.

The shift also reframes learning. Learning does not have to be complete to be useful. Women can understand basic properties without mastering technical detail. They can stop and resume. They can decide that understanding is sufficient without acting. Knowledge does not obligate use. This matters because attention is scarce. Women often manage multiple demands simultaneously. A system that requires constant vigilance or immersion competes with essential responsibilities. Bitcoin allows intermittent engagement. It does not punish pauses. That flexibility preserves capacity.

Critical distance is also restored. Women can question claims without being accused of disloyalty. They can acknowledge limits without dismissing value. They can engage skeptically without disengaging entirely. A tool that survives scrutiny is more reliable than one protected by narrative.

This distance supports better risk assessment. Women can see where Bitcoin introduces new exposure and where it reduces existing risk. Custody responsibility, volatility, and legal context remain visible. Decisions become proportional rather than reactive. Calm replaces urgency.

Choosing how to use Bitcoin also means choosing boundaries. Women can decide how much exposure is appropriate. How often to engage. Under what conditions to stop. These boundaries prevent drift. They protect autonomy by keeping choice active rather than assumed.

The shift also restores the right to quiet experimentation. Women do not owe explanation to institutions, communities, or critics. Financial exploration can remain private. Privacy here is not secrecy. It is space to learn without pressure or performance. This quiet approach is especially important in tightening environments. When systems increase monitoring and compliance, discretion becomes protective. Bitcoin can be explored without signaling alignment or dissent. Choice remains tactical rather than symbolic.

The shift also acknowledges change over time. Needs evolve. Risk tolerance shifts. Environments tighten or loosen. What is unnecessary now may matter later. Bitcoin as an option accommodates this movement. Identity framing does not. Women can return to the tool when conditions change without having betrayed a position.

Responsibility is re centered appropriately. Bitcoin does not remove responsibility. It clarifies where it resides. Women choose when to accept control and when to accept delegation. They choose which risks to carry and which to avoid. That clarity reduces confusion and resentment.

This chapter does not argue for a specific outcome. It argues for clean decision making. Bitcoin does not need to be embraced or rejected wholesale. It needs to be understood well enough to be used deliberately or declined confidently.

The shift is subtle but decisive. When Bitcoin is treated as an option, women regain authority over timing, scope, and exposure. They can integrate it where it strengthens autonomy and leave it aside where it does not. Choosing when and how to use Bitcoin is not about predicting collapse or defending a future. It is about responding to present constraints with tools that fit. In a landscape where control is often centralized and permission is increasingly conditional, having an option that can be chosen quietly matters. That ability to choose without declaration is the shift.

Conclusion

This book does not ask women to believe in Bitcoin. It asks them to understand it well enough to decide where it belongs in their lives. Across these chapters, the focus remains consistent. Systems shape behavior. Permission determines access. Control matters most when pressure appears. These realities are not abstract. They surface in moments of urgency, care, responsibility, and constraint.

Bitcoin is examined as infrastructure, not as a promise. Infrastructure has limits. It carries costs. It solves some problems and leaves others untouched. Seeing it clearly allows women to use it without illusion or fear. That clarity restores choice in environments where choice is narrowed by design.

Control rarely arrives announced. It expands quietly through systems that manage money, identity, and compliance. When those systems tighten, women are expected to absorb the burden. More vigilance. More explanation. More emotional labor. Understanding where control resides allows resistance without collapse.

Bitcoin does not remove risk. It rearranges it. It creates options, not safety. Options matter when timing, discretion, and independence are at stake. Discernment is required here. In a world where permission is increasingly conditional and visibility increasingly enforced, having the ability to choose quietly is power. That power does not require belief. It requires understanding.

About the author

The author lives removed.

Please feel free to burn part or all of this book, safely, as
an effigy.

www.ingramcontent.com/pod-product-compliance
Lightning Source LLC
Chambersburg PA
CBHW020943090426
42736CB00010B/1245